Modeling and Control of Physical Systems

Dr. Kyihwan Park

GIST PRESS
028

Modeling and Control of Physical Systems

Dr. Kyihwan Park

GIST PRESS
광주과학기술원

"To my honorable parent, Park Kwan-yong, Jung Eun-hee,
beloved wife, Jae-Kyung Sohn,
proud sons, Jae-Young, Joo-Hyung, Jae-Won Park,
and dear all my students"

Preface

Modern high-tech engineering requires a study of multi-disciplinary research for integrated systems. For this, it is essential to have a wide knowledge and understanding of the physical system dynamics of mechanical, electrical, magnetic, hydraulic, and pneumatic systems. To understand the concept of energy or power exchange of various physical systems or plants, a bond graph modeling technique is introduced in Chapters 2 and 3 as a universal approach for physical system (plant) modeling.

The physical systems can be understood using the mathematical description of the systems in terms of differential equations described in the time domain. However, it is sometimes convenient to interpret a system in the frequency domain, particularly when it is required to determine the relation between input and output, which is called a transfer function. It is covered in detail in Chapter 4. Additionally, understanding the relation between system dynamics and performance in terms of speed and precision will be addressed.

Most of the physical systems are driven using electrical power, such as voltage and current. Therefore, it is essential to comprehend the principles of an electromechanical system, such as an electrical motor or actuator, which helps clarify the conversion process of electrical power into mechanical power. Furthermore, the study of magnetic systems is necessary because electrical power is transformed into mechanical power through magnetic systems. These types of power transformation are covered in Chapter 5.

A good mechanical design should strive to have good system characteristics in terms of speed and accuracy because it can perform its tasks without relying on controllers or compensators. However, it is not always easy to design a plant that realizes the expected performance. Sometimes mechanical systems have some components that are unstable or have a small damping ratio that causes large vibrations. Also, uncertainties exist in the high-frequency region, which makes system modeling difficult. These types of problems are expected to be solved by using feedback controllers. Chapter 6 introduces traditional control theory with more emphasis on frequency analysis. Several applications in real systems are also given as examples.

The actual implementation of a feedback controller is important to organize a closed-loop system. Conventional controllers, such as proportional-integral-derivative (PID) controllers or compensators, can be easily implemented using analog electronic circuits, thus benefiting from the ease of connection to an analog plant (system). Chapter 7 introduces how to construct a feedback controller using analog electronic circuits using operational amplifiers. We can also implement a mechanical plant using electronic circuits. In combination, feedback control simulations are performed experimentally to investigate the control results before a mechanical plant is actually controlled. Some of the electronic components are also introduced for real feedback control implementation.

Another essential component of the feedback system is a sensor that is used for measuring the current state of the plant, such as position and velocity. Fundamental signal processing techniques, especially for an optical sensor, are introduced to improve performances in terms of bandwidth and resolution in Chapter 8.

Contents

Conventional modeling methods

—

Conventional modeling methods

1.1   Modeling process

Let me start with a question before answering why modeling is important. An object is falling on a wooden table with legs, as shown in Figure 1.1, and you are asked to describe the behavior of the wooden table in terms of displacement or velocity in the x direction. What would be your first step, and what process is required to answer this question?

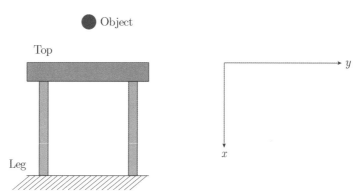

Figure 1.1: A schematic diagram of an object falling on a wooden table (left) and the coordinates chosen for describing its behavior (right).

There are several aspects to consider when solving this question, including what constitutes the table, and determining the thickness and flexibility of table legs. It is also important to determine the types of inputs acting on the table. We must consider whether they should be regarded as force or velocity. Moreover, a mathematical approach should be selected to describe the behavior of the wooden table.

The purpose of the modeling process is to ultimately achieve realistic and approximate step-by-step solutions for the aforementioned aspects. If the table has a heavy top and flexible legs, it is treated as a mass and a spring connected serially with the force input, as shown in Figure 1.2(a). If the top is light and rigid, it can be treated as mass-less. Then, the table can be modeled with only a spring with a velocity input, as shown in Figure 1.2(b). Figures 1.2(a) and 1.2(b) present schematic models of the wooden table corresponding to the different cases. Schematic diagrams are well known for their effectiveness in providing graphical representation for visualizing the operation and function of physical systems.

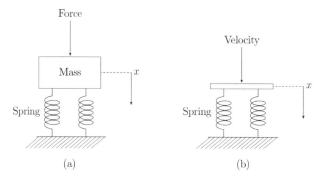

Figure 1.2: Schematic diagrams constructed based on different assumptions: (a) heavy and (b) mass-less tabletop with flexible legs.

As observed in the different schematic diagrams, the actual configurations of the structure and system must be properly visualized in the model for the design purpose. For example, if the wooden table is used for vibration reduction, it is more appropriate to model the table with a mass, as shown in Figure 1.2(a) because mass is related to vibration characteristics. A mathematical description of the model is then obtained analytically. This model analysis can be used to change the system parameters, such as table mass and leg stiffness, to achieve the best system behavior performance. Modeling can also help ensure that the system is optimally designed for its application purpose.

Q1: Sketch the displacement and velocity of the top of the table in the x direction with respect to time.

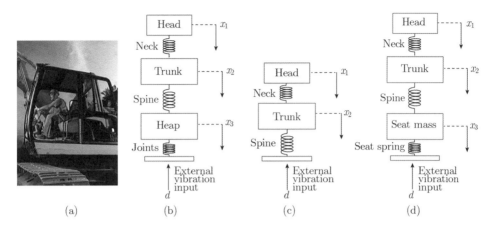

Figure 1.3: (a) Actual configuration of an excavator (https:// www.123rf.com /photo 32442180, a worker operating an excavator on construction site.html), (b), and (c) different schematic diagrams of the person's body, and (d) the final model for the seat design.

Here is another example: suppose that you are developing an excavator seat, similar to the one shown in Figure 1.3(a) to reduce neck injuries by transmitting external vibration through the seat from the ground [1]. Assuming the head of the person in the seat is a rigid body, a model can be designed such that the head is a mass and the neck is a mass-less spring. The trunk, spine, heap, and joints can be assumed and modeled as rigid bodies having masses and springs, as shown in Figure 1.3(b). The body of a person can be systematically analyzed while considering head, trunk, and heap displacements x_1, x_2, and x_3, respectively for an external vibration d applied from the ground through the seat. However, this model can be much simplified, as shown in Figure 1.3(c) by combining the heap and joints, which then will be ignored due to the higher stiffness of joints compared to that of the neck and spine. Furthermore, they can be regarded as moving together with the seat. Of course, the model depicted in Figure 1.3(b) can be more accurate even though its analysis is more complicated than that of Figure 1.3(c). However, a simple model design is preferred when the results are similar because it not only saves time in analysis but also delivers concepts more clearly. When the seat is also modeled using a mass and spring, we can achieve the final analysis of seat design by adding them to the person's body modeling as shown in Figure 1.3(d).

1.2 Input sources in mechanical systems

We start with discussing input sources of a mechanical system for its schematic modeling as its dynamic behavior occurs by the input, which is known information. It's important to choose an appropriate input because choosing the wrong one will make system analysis difficult. Eventually, the system may fail unexpectedly, leading to unpredictable results.

Force and velocity are two types of input sources to a mechanical plant or system. They are generally called effort and flow sources. When there is a mass on the surface, as shown in Figure 1.4, which input along with force $f(t)$ and velocity $v(t)$ is applied to the mass? Suppose that $f(t)$ is applied to the mass m; $v(t)$ is obtained by integrating $\frac{f(t)}{m}$ and displacement is obtained by integrating $\frac{f(t)}{m}$ twice. Meanwhile, what if $v(t)$ is directly applied to the mass? Is this a proper assumption? Is it necessary to assume that the mass exists even when the known velocity input $v(t)$ is applied? The answer is no because $v(t)$ is applied regardless of the mass existence. Accordingly, $f(t)$ would be the appropriate input to the mass that reflects the intention of a designer.

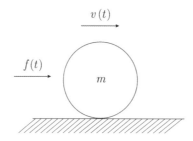

Figure 1.4: Force input to a mass.

What about the case shown in Figure 1.5, where the input is applied to a spring attached to the wall? Between $f(t)$ and $v(t)$, which would be the proper input? Suppose that $v(t)$ is applied to the spring. Then, the spring force, $f(t)$, is obtained using the relative displacement of the spring calculated by integrating the relative velocity. Meanwhile, suppose that $f(t)$ is directly applied to the spring. Is this the correct assumption? Do we need to assume the spring exists even when the known $f(t)$ is applied? The answer is no because the force is already computed by considering the spring. Accordingly, velocity would be the appropriate input in this case, reflecting the intention of a designer.

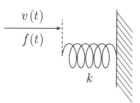

Figure 1.5: Velocity input to a spring.

Another thumb rule to determine an input source is by checking whether the motion occurs in contact or without contact with an input source. When mechanical motion occurs in contact with an input, the source is generally a displacement (velocity). When mechanical motion occurs without contact with an input, the source is generally a force. One example of a force input can be found in the gravity force. It is considered an ideal force input because it is always constant, regardless of velocity change. Another example is an electrical direct current (DC) motor because force or torque is generated by the electromagnetic principle without contact.

However, the above thumb rule is not always correct. When the mass is generally small, velocity can be an input even if motion occurs in a non-contact way. One example of a velocity input with no contact motion can be found in a step motor, where the angular velocity of a rotor is determined by the switching frequency of the pole magnetization. In this case, the mass of the rotor is assumed to be negligibly small so that it does not affect the velocity input. In this case, the velocity input can be considered ideal. However, when the mass of the rotor increases, the velocity input cannot be considered an ideal input any longer because the rotor may have slip motion due to the heavy mass. Hence, the load condition, such as the magnitude of a mass, should also be considered when determining the input types.

Q2: Figure 1.6 shows an inch-worm motor driven by a piezoelectric actuator whose displacement is proportional to the applied voltage [2]. The rod in the center can move like a bug crawling by compressing, extending, and releasing the piezoelectric actuators on the sides, top, and bottom in the order of Step 1 to Step 5. Which input source modeling is more appropriate for the inch-worm motor? Suppose the mass of the rod is $100\,gr$, The compressing force of the actuator on the rod is $10\,N$, the friction coefficient is 0.5, and the gripping and

detaching frequencies are $100\,Hz$. How fast is the rod moving forward when one step moving distance is $0.1\,mm$?

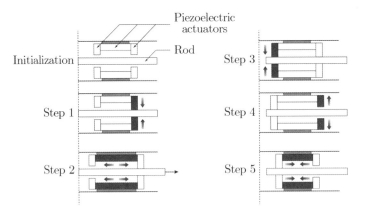

Figure 1.6: An inch-worm motor.

Q3: Figures 1.7(a) and (b) show the configuration of a DC motor and a step motor. A DC motor works by flowing electric current through an electromagnet (rotor) in a magnetic field generated by a permanent magnet. A step motor rotates a predetermined step angle of a rotor by the number of input pulses applied to the stator. What is the difference between these motors in terms of input source type?

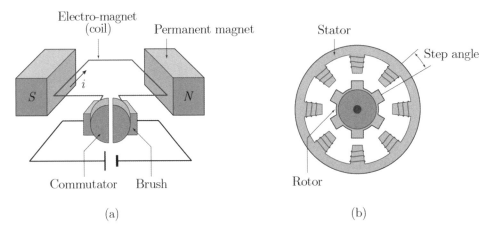

Figure 1.7: Configurations of (a) a DC motor and (b) a step motor.

1.3   Constitutive relations

There are various engineering systems such as mechanical, electrical, and fluid systems with graphical definitions, as shown in Figures 1.8(a), (b), and (c). Here, f and v are the applied force and velocity, respectively, in a mechanical system comprising composed of mass m, spring k, and damper c. V and i are the applied voltage and current, respectively, in the electrical circuit composed of an inductor L, resistor R, and capacitor C. P_1, P_2, and Q are the inlet pressure, outlet pressure, and flow rate, respectively in the fluid system composed of an area-adjustable valve R and flexible pipe C. Their governing equations are obtained by applying the well-known principles of physics, such as Newton's law, Kirchhoff's circuit law, and Bernoulli equation, as indicated in Equations (1.1), (1.2), and (1.3), respectively.

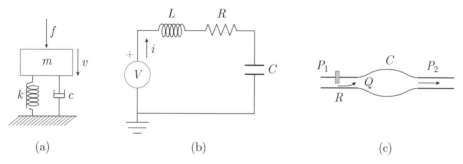

(a)   (b)   (c)

Figure 1.8: Several physical systems represented using schematic diagrams.

- Newton equation (mechanical system)

$$f = m\ddot{x} + c\dot{x} + kx \tag{1.1}$$

- Circuit equation (electrical system)

$$V = L\frac{di}{dt} + Ri + \frac{1}{C}\int i\,dt \tag{1.2}$$

- Bernoulli equation (fluid system)

$$P = RQ + \frac{1}{C}\int Q\,dt \tag{1.3}$$

These types of equations are usually used for system analysis. It should be noted that there are typical variables and parameters in different physical systems. It is important to ask if there are any similarities between the typical variables and parameters in terms of physical concepts. If any such similarities exist, we can understand other systems through a unified approach applied to a specific system. For example, if there are variables and parameters corresponding to force/velocity and mass/spring/damper of a mechanical system in other electrical, magnetic, fluid, and thermal systems, it would be easy to understand other analogous systems based on the knowledge and physical concept associated with one field. Moreover, we do not need to memorize the equations related to different systems.

In the next chapter, we will find out the similarities between the physical systems using a graphical unified modeling approach called a bond graph in more detail. The similarities between them are first observed by means of constitutive relations.

1.3.1 Variables and parameters in a mechanical system

There are four variables in a mechanical system: force f, velocity v, momentum p, and displacement x. f and v are called power variables because the product of f and v is power P. We have relations of p and x from their definitions.

$$f = \frac{dp}{dt}, \quad v = \frac{dx}{dt} \tag{1.4}$$

Hence, only two variables among the four are independent of each other.

The mechanical energy, E, is defined by integrating P. Thus, the potential energy E_p and kinetic energy E_k can be differently represented using Equation (1.4) as

$$E_p = \int P dt = \int f v dt = \int f \frac{dx}{dt} dt = \int f dx \tag{1.5}$$

$$E_k = \int P dt = \int v f dt = \int v \frac{dp}{dt} dt = \int v dp \tag{1.6}$$

p and x are called energy variables as they appear in the energy relations indicated in Equations (1.5) and (1.6).

Moreover, there are similar variables equivalent to f, v, p, and x in other physical systems such as electrical, fluidic, magnetic, and thermal systems. We introduce the generalized variables effort (e), flow (f), momentum (p), and displacement (q) to represent the variables of all physical systems. The definitions made for the variables represented in a mechanical system using Equation (1.4) can be also applicable to other systems. We recognize that there are similarities among variables in those physical systems, as listed in Table 1.1 [3].

Table 1.1: Variables in physical systems.

Physical systems	effort (e)	flow (f)	momentum (p)	displacement (q)
Mechanical system (linear)	f	v	p	x
Mechanical system (rotational)	T	ω	H	θ
Electrical system	e	i	λ	q
Hydraulic system	P	Q	Γ	V
Magnetic system	M	$\dot{\phi}$		ϕ

There are only three components in mechanical systems, mass m, damper characterized by a damping coefficient c, and spring represented by spring coefficient k. The three parameters, m, c, and k can be represented using the four variables f, v, p, and x, which are called constitutive relations. When spring and damping forces are indicated as f_s and f_s, respectively, the constitutive relations are as follows:

$$v = \frac{p}{m} \tag{1.7}$$

$$f_s = kx \tag{1.8}$$

$$f_d = cv \tag{1.9}$$

In the above equations, m is a linear parameter that relates momentum to velocity. k is a linear parameter that relates force to displacement. c is a linear parameter that relates force to velocity. The above constitutive relations are graphically drawn in Figure 1.9. Equations (1.7), (1.8), and (1.9) are called linear constitutive relations. These constitutive relations are also held in other physical systems. In some systems, they can be nonlinear such as in a fluid system.

Using the linear constitutive relations, Equations (1.5) and (1.6) are respectively represented as follows:

$$E_p = \int f_s dx = \int kx dx = \frac{1}{2}kx^2 \tag{1.10}$$

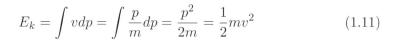

$$E_k = \int v\,dp = \int \frac{p}{m}\,dp = \frac{p^2}{2m} = \frac{1}{2}mv^2 \tag{1.11}$$

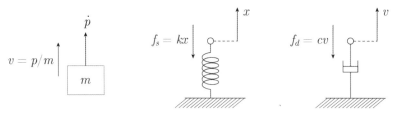

Figure 1.9: Constitutive relations in a mechanical system.

1.3.2 Variables and parameters in an electrical system

There are four variables in an electrical system: voltage e, current i, momentum λ, and displacement q, as listed in Table 1.1. e and i are called power variables because the product of e and i is a power. λ and q are called energy variables.

Moreover, there are also only three parameters in an electrical system, which are inductor L, resistor R, and capacitor C like a mechanical system. The electrical momentum λ may be unfamiliar to you. Even though it will be again introduced in Chapter 5, here, it is simply defined as a measure of how well a magnetic flux couples with that of another coil. For example, the more turns of an electric coil through which current flows, the higher λ is because a higher magnetic flux is produced through the coils.

Similarly, the definitions made for the mechanical system represented in Equations (1.7), (1.8), and (1.9) are also applicable to electrical systems where the corresponding constitutive relations are

$$\lambda = Li_L \tag{1.12}$$
$$e_C = \frac{1}{C}q \tag{1.13}$$
$$e_R = Ri_R \tag{1.14}$$

where L can be defined as a parameter that relates flux linkage (momentum) λ to current (flow) i_L flowing through the inductor. C can be defined as a parameter that relates voltage (effort) e_c to charge (displacement) q. R can be defined as a parameter that relates voltage (effort) e_R to current (flow) i_R flowing through the resistor R.

We also have a relation of e_L derived from the definition that effort is the time derivative of the momentum, i.e.,

$$e_L = \frac{d\lambda}{dt} \tag{1.15}$$

In addition, we have a relation for i_C derived from the definition that flow is the time derivative of the displacement, i.e.,

$$i_c = \frac{dq}{dt} \tag{1.16}$$

Hence, using Equations (1.12) and (1.13), e_L and e_C can be rewritten as

$$e_L = L\frac{di_L}{dt} \tag{1.17}$$

$$e_C = \frac{1}{C}\int i_C dt \tag{1.18}$$

The above constitutive relations are graphically drawn in Figure 1.10.

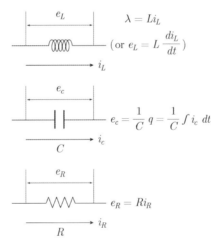

Figure 1.10: Constitutive relations in an electrical system.

1.4   Modeling of mechanical systems

Now, we will apply the above modeling process to a real mechanical system. Figure 1.11(a) shows a cam and follower that is used for a gas combustion valve in an automobile engine. Figure 1.11(b) shows its corresponding schematic diagram. Mass m_f is modeled to represent the rigid body of the follower and valve. Spring, k_2 is modeled to represent the stiffness of the roller moving up and down in contact with the cam. Spring k_1 is modeled to represent the stiffness of the mechanical spring located between the fixed body and the moving body attached to the follower. Damper c is also modeled to represent the friction associated with the component motion. x_c and x_f are the displacements of the cam and follower, respectively. The displacement (or velocity) of the cam x_c is selected as an input because the mechanical motion occurs in a contact way. Then, the governing equation is derived using Newton's law as follows:

$$m_f\ddot{x}_f + c\dot{x}_f + (k_1 + k_2)x_f = k_2 x_c \tag{1.19}$$

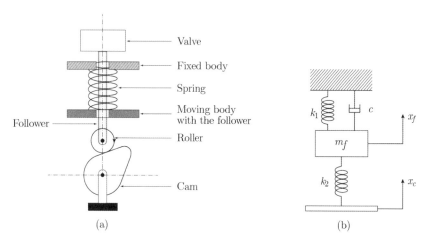

(a) (b)

Figure 1.11: (a) Cam and follower configuration and (b) its corresponding schematic diagram.

The parameters m_f, k_1, and k_2 are chosen to fulfill the design purpose, which is to make x_f follow x_c as closely as possible, i.e., $x_f \cong x_c$. This requires solving the second-order differential equation described using Equation (1.19) to find out how they are determined. Time and frequency domain analyses will be used to determine an accurate solution in Chapter 4. If we simply assume that m_f and c_f are negligibly small, Equation (1.19) is expressed as

$$(k_1 + k_2)x_f = k_2 x_c \tag{1.20}$$

From Equation (1.20), a large k_2 and small k_1 are selected to achieve the desired result of $x_f \cong x_c$.

In Figure 1.11(b), the displacement (or velocity input) x_c has been selected as the input. However, as you can see, it would seem that force could also be used as an input, as Equation (1.19) can be transformed to

$$m_f \ddot{x}_f + c\dot{x}_f + k_1 x_f = k_2(x_c - x_f) = f \qquad (1.21)$$

The corresponding schematic diagram is presented in Figure 1.12. Equation (1.21) appears to be an equivalent equation with force input f, which is the same as $k_2(x_c - x_f)$. However, f is not obtained without knowing x_f which will be obtained as a result when f is applied. Hence, there is a conflict that the output x_f should be known independently to determine the input f. Therefore, it can be said that the displacement input is a right choice in this case. This result agrees with the fact that the velocity input is applied to a spring, not a mass. Similarly, the force input is applied to a mass, not a spring.

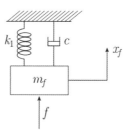

Figure 1.12: A schematic diagram of the cam-follower device equivalently modeled with a force input.

As another example, Figure 1.13(a) shows a 1/4 scale vehicle suspension that is used for absorbing the impact on an automobile body from a rough road. Figure 1.13(b) shows the corresponding schematic diagram. Here, m_v and m_t are the masses of the vehicle and tire, respectively. k_1 and k_2 are the stiffnesses of the real mechanical spring in the absorber and tire, respectively. c is the damping coefficient of the absorber. x_2, x_1, and x are the displacements of the vehicle, tire, and input from the ground, respectively. The governing equation is derived using Newton's law with a displacement input x as follows:

$$m_t \ddot{x}_1 + c(\dot{x}_1 - \dot{x}_2) + k_1(x_1 - x_2) + k_2 x_1 = k_2 x \qquad (1.22)$$
$$m_v \ddot{x}_2 + c(\dot{x}_2 - \dot{x}_1) + k_1(x_2 - x_1) = 0 \qquad (1.23)$$

We can use Equations (1.22) and (1.23) for the vehicle suspension analysis. The

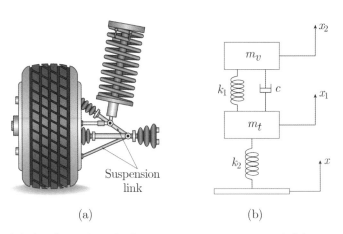

(a)   (b)

Figure 1.13: (a) A 1/4 scale vehicle suspension system and (b) its corresponding schematic diagram.

design purpose of this device is to suppress x_2 for x input as much as possible, such that $x_2 \cong 0$.

The displacement x is chosen as an input because it is applied directly with contact to spring k_2. Furthermore, displacement information is more easily available from the road surface than force information because displacement information can be described by a mathematical representation of the road surface. By the way, if we move the $k_2 x_1$ term on the left side of Equation (1.22) to the right, we can transform it into an equivalent equation with force input, $k_2(x - x_1)$ to the mass m_t. However, this model is not practical because the force is not obtained without knowing x_1 and x as similarly stated in the above example. Therefore, it can be said that the displacement input is a right choice in this case. In conclusion, the modeling procedure for mechanical systems is summarized as follows:

1. First, input types should be selected, considering whether the input is applied to a mass or a spring. The input to a mass should be the force while that to a spring should be the velocity. One easy way to determine an input source is by checking whether motion occurs in contact or non-contact with the input source. When mechanical motion occurs in contact with an input, the source is displacement (velocity). When mechanical motion occurs without contact with the input source, the source is a force.

2. The input type should be selected such that it is more convenient to obtain the governing equations.

3. Correct assumptions should be made for the mass, spring, and damper to satisfy the design purposes.

4. Modeling should be as simple as possible to save analysis time and to deliver a clear physical concept.

Q4: Figure 1.14 shows an earthquake seismograph that is used for recording vibration signals due to earthquakes. Draw its schematic diagram and derive the governing equation. What is the design criterion? Particularly, what should the output motion be for the input vibration?

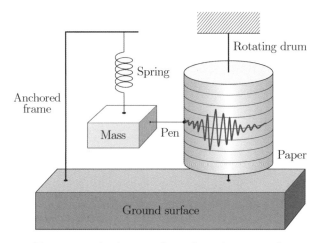

Figure 1.14: An earthquake seismograph.

Q5: Figure 1.15 shows a door closer that provides a restoring force to close a door when it is open. Draw its schematic diagram and derive its governing equation.

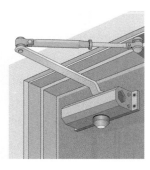

Figure 1.15: Picture of a door closer.

1.5 Load effect

1.5.1 The load curve of a mechanical system

If a load is defined as any type of component that reduces the magnitude of an effort source or a flow source, it is important to recognize that the three parameters of a mechanical system—mass, spring, and damper—are considered loads because they work to reduce the input power. Therefore, the applied input power should be sufficient to provide force and velocity to the loads at the final stage as desired.

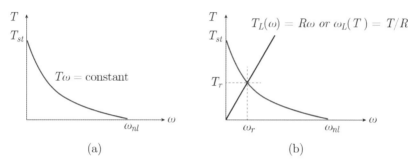

(a)   (b)

Figure 1.16: (a) T versus angular velocity ω curve of the automobile engine and (b) the final resultant torque and angular velocity obtained by $T - \omega$ and the load curves.

Another thing to note here regarding the input power is that the force and velocity inputs can be changed depending on the load conditions. As an example, Figure 1.16(a) shows a typical torque, T curve for an angular velocity, of an automobile engine, where the product of T and ω is constant, assuming constant gas supply. Here, T_{st} is the stall torque, which is the maximum torque obtained at zero angular velocity. ω_{nl} is the no-load angular velocity, which is the maximum angular velocity obtained at zero torque. Suppose that the loads of the automobile, such as the mass of the body, slope of the road, and friction of the tire, are modeled using equivalent resistance R. Then, the torque generated by the loads, T_L is expressed using the constitutive relation $T_L = R\omega$ referring to the rotational mechanical system presented in Table 1.1. When the automobile is in motion, T_{st} decreases as ω increases. Resultant torque T_r (real torque) and resultant angular velocity ω_r (real velocity) can be determined using the load curve $T_L = R\omega$ and $T - \omega$ curve ($T\omega$=constant), as shown in Figure 1.16(b).

In this case, T_r is not constant but changed depending on the load torque T_L that varies with ω. Similarly, if the flow ω_{nl} is an input, ω_{nl} is decreased due to the loads mentioned above. Resultant angular velocity ω_r and resultant torque T_r are determined by the load curve specified by $\omega_r = T/R$ and constant power $T - \omega$ curve ($T\omega$=constant).

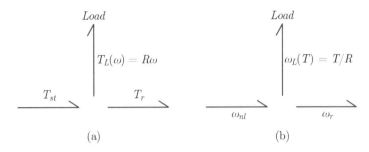

(a)   (b)

Figure 1.17: Illustration of (a) real torque input and (b) real angular velocity input.

The above statement for determining T_r and ω_r is depicted using the half arrow bonds, as shown in Figures 1.17(a) and (b). These indicate that the constant inputs T_{st} and ω_{nl} can be represented by load-dependent real inputs T_r and ω_r. T_r and ω_r are mathematically represented as

$$T_r \;\; = \;\; T_{st} - T_L(\omega) = T_{st} - R\omega \tag{1.24}$$
$$\omega_r \;\; = \;\; \omega_{nl} - \omega_L(T) = \omega_{nl} - T/R \tag{1.25}$$

As discussed above, it is difficult to supply a constant force or velocity to a system because of the load on the system. However, there are many cases where you need to supply a constant force or velocity to a mechanical system regardless of the load. How do we accomplish this? The answer can be found in Chapter 6.

Q6: If the torque T_r is not sufficient for an automobile to climb an uphill road how should the load curve be changed? How is a change in the load curve possible?

1.5.2   The load curve of an electrical system

The effect of a load in a physical system is easily understood using an electric circuit model. Suppose the overall load is R for the circuit shown in Figure

1.18(a), which has a constant voltage input, e_{in}, that is realized using voltage control. Then, the current i_R due to the overall load is obtained using the electrical relation: $i_R = e_{in}/R$. This result is also graphically determined using the load curve specified as $e_R = Ri_R$ and the constant voltage condition when e_R is defined as the voltage across the load. For example, when $e_{in} = 10\,V$, $R = 1\,k\Omega$, e_R and i_R are determined as $10\,V$ and $0.01\,A$, respectively, based on the graph shown in Figure 1.18(b).

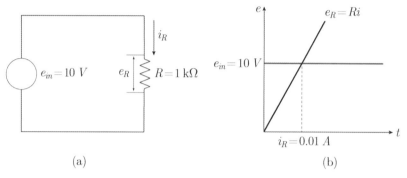

(a)   (b)

Figure 1.18: (a) Load modeled using a resistor and (b) load curve to determine current.

When another load is added in parallel to the initial circuit, as shown in Figure 1.19(a), it is possible to calculate the loads using an equivalent load or resistance R_{eq}. The current is newly obtained from the electrical relation of $i = e_{in}/R_{eq}$. Similarly, current due to the loads is also graphically determined using the load curve specified as $e_R = R_{eq}i$ with a constant voltage condition. Because its slope is less steep than that of the single resistor, e_R and i_R are determined as $10\,V$ and $0.02\,A$, respectively, based on the graph shown in Figure 1.19(b).

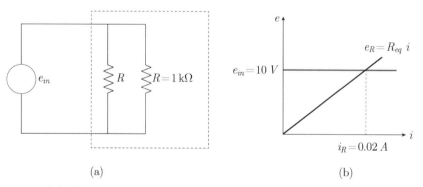

(a)   (b)

Figure 1.19: (a) An electric circuit with another load added in parallel and (b) its corresponding load curve to determine current.

1.6 Passive system; non-isolated system

Suppose there is a mechanical system composed of mass (m_1), damper (c_1), and spring (k_1) with force input, as shown in Figure 1.20(a). The governing equation of this system is obtained and described using a second-order differential equation. If another mechanical system composed of mass (m_2), damper (c_2), and spring (k_2) is serially connected to the previous system to have a two-degree-of-freedom motion, as shown in Figure 1.20(b), can each system be described separately using the mechanical equation obtained from the system shown Figure 1.20(a)?

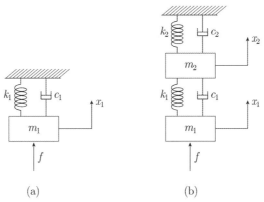

(a)   (b)

Figure 1.20: A schematic diagram of a mechanical system with (a) one and (b) two-degree-of-freedom.

We first derive the equation of the mechanical system shown in Figure 1.20(a). We obtain

$$f = m_1\ddot{x}_1 + c_1\dot{x}_1 + k_1 x$$

The equations of the mechanical system shown in Figure 1.20(b) are represented as

$$f + c_1\dot{x}_2 + k_1 x_2 = m_1\ddot{x}_1 + c_1\dot{x}_1 + k_1 x \qquad (1.26)$$

$$m_2\ddot{x}_2 + (c_1 + c_2)\dot{x}_2 + (k_1 + k_2)x_2 = c_1\dot{x}_1 + k_1 x_1 \qquad (1.27)$$

Using Laplace transformation [4], we can easily obtain the relation between the input force and output displacement from the above equations. However, this step is deferred until the response analysis is studied in Chapter 4. Instead, the differential equations are used to determine the input and output relations. As written in Equations (1.26) and (1.27), x_1 is coupled with x_2 and parameters,

c_2, k_2, c_1, and k_1. Because the trailing part is coupled with the previous part, the governing equation obtained for the previous part shown in Figure 1.20(a) is no longer valid. The governing equation should be derived again from the beginning to investigate the dynamic behavior of the output displacement, x_1. In conclusion, it is impossible to describe each system separately like two isolated systems.

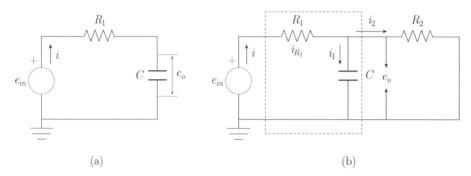

(a)   (b)

Figure 1.21: (a) An electronic circuit composed of R_1 and C and (b) another circuit serially connected to the previous circuit with a resistor R_2.

We repeat this question for the electronic circuit shown in Figure 1.21(a), which is composed of resistor R_1 and capacitor C with voltage input e_{in} and output voltage e_o across C. If another circuit composed of resistor R_2 is serially connected to the previous circuit, as shown in Figure 1.21(b), is it possible to obtain the relation of $\frac{\text{output voltage } e_o}{\text{input voltage } e_{in}}$ without considering R_2? This question can be answered by deriving an electrical equation of the electrical circuit. However, it can be similarly understood through the analysis conducted for the mechanical system. It is observed that e_o is not only dependent on R_1 and C but also dependent on R_2 because current i across C in the circuit shown in Figure 1.21(a) divides into i_1 and i_2 in the circuit shown in Figure 1.21(b). In other words, the previous circuit is affected by the next adjacent part of the circuit, which works as a load on the previous circuit. In conclusion, it is unavoidable to have a non-isolated system if a system is passive. To have an isolated system that is not affected by the loads, it must be an active system, such as a feedback system (servo system). A classic example can be found in electronic operational amplifiers where a feedback loop is implemented, which is introduced in Chapter 6.

Bond graph modeling for mechanical systems

Bond graph modeling for mechanical systems

2.1   Power flow in physical systems

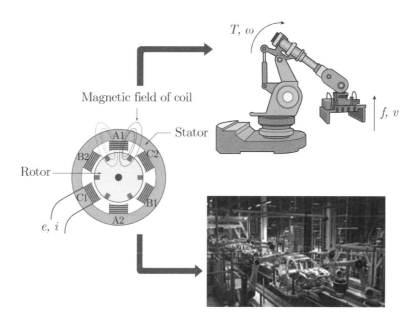

Figure 2.1: Applied electrical power delivered to the mechanical plants in an industrial robot.

It was stated in Chapter 1 that there are power and energy variables, namely, effort, flow, momentum, and displacement, in mechanical and electrical systems. Figure 2.1 shows one of examples showing how the applied electrical power ei is delivered to the mechanical plants in an industrial robot, converted to $T\omega$ and fv. When the electrical motor generates the current i from the applied voltage, the generated electromechanical torque T is applied to the robot arm. Then, it rotates the robot arm at the angular velocity ω. The translated velocity v from ω lifts the mass. The power flow is simply represented using the conventional block diagrams in Figure 2.2.

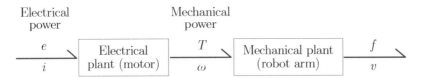

Figure 2.2: Electrical power flow of an industrial robot represented using block diagrams.

In the above statement, if the input power is not sufficiently applied, the weight will not be lifted with the desired velocity. Therefore, the applied input must be well-designed to meet the desired performance. However, in the real world, it is common to encounter power loss and storage during power transmission through several plant components due to power-dissipated components or storing loads. Hence, it is necessary to figure out how power or energy is delivered to the plant components.

However, the block diagram shown in Figure 2.2 is not an appropriate tool to investigate the above analysis because the system is oversimplified in the block diagram representation. Thus, a modeling process is preferred that shows the transmission of power or energy through an individual component or load composing the plant from the input source. Furthermore, it would be desirable to provide accurate and detailed information about the plant components, their modeling process, and plant designers' intentions. This is why the bond graph modeling technique [3],[5],[6] is introduced.

2.2   Bond graph modeling in mechanical systems

It was stated in Chapter 1 that there are only three components in mechanical systems: mass m, a damper represented by damping coefficient c, and a spring represented by spring coefficient k. There are no other elements except these three elements. There are four variables: force f, velocity v, momentum p, and displacement x. f and v are called power variables. p and x are called energy variables.

2.2.1 Passive 1-ports: m, c, k

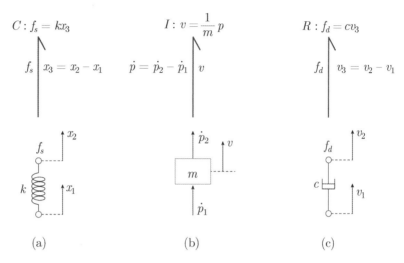

Figure 2.3: Passive 1-ports of a mechanical system.

When displacements x_1 and x_2 are applied as inputs to the spring as shown in Figure 2.3(a), energy is transmitted to the spring and potential energy is stored. Then, spring force, f_s is represented using the following constitutive relation:

$$f_s = kx_3 = k(x_1 - x_2) \tag{2.1}$$

where k is the spring stiffness. The energy transmission to the spring element is represented using a passive 1-port with C element. The reason it is called a passive 1-port is because it exchanges power (or energy) at a single location or port and contains no sources of power. The bond with a half arrow ($\rightharpoonup$) is used to indicate the direction of power or energy.

When forces expressed by $\dot{p}_1$ and $\dot{p}_2$ are applied to mass m, as shown in Figure 2.3(b), energy is transmitted to the mass and kinetic energy is stored. Then, velocity v is represented using the following constitutive relation:

$$v = \frac{1}{m}\Delta p = \frac{1}{m}(p_1 - p_2) \tag{2.2}$$

The energy transmission to the mass element is represented as a passive 1-port with an I element. The bond with a half arrow is also used to indicate the direction of power or energy.

Lastly, when velocity v_1 and v_2 are applied to the damper, as shown in Figure 2.3(c), the power is transmitted to the damper and is dissipated. Then, the damping force f_d is represented using the following constitutive relation:

$$f_d = cv_3 = c(v_1 - v_2) \tag{2.3}$$

where c is the damping coefficient. This power transmission to the damping element is represented as a passive 1-port with an R element. The bond with a half arrow is used to indicate the direction of power.

2.2.2 Junctions

1-junction (common velocity junction)

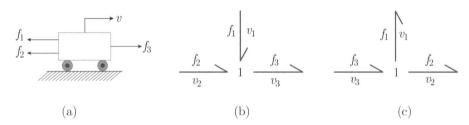

(a)   (b)   (c)

Figure 2.4: (a) A cart on rollers where three forces are acting, (b) 3 port 1-junction diagram, and (c) another 3 port 1-junction diagram.

Figure 2.4(a) shows the schematic diagram of a mechanical system that looks similar to a cart on rollers where three applied forces, f_1, f_2, and f_3 are acting. The cart is assumed to be mass-less. The forces in the free-body diagram shown in Figure 2.4(a) are labeled in the three bonds around the 1-junction with velocities, as shown in Figures 2.4(b) or (c) to indicate that the velocity of the three points of force acting on the cart is identical.

$$v_1 = v_2 = v_3 = v \tag{2.4}$$

where v is a common velocity. Hence, the 1-junction is called a common velocity junction. By the way, because the cart is mass-less, the force relations of the free-body diagram are obtained as

$$f_1 + f_2 - f_3 \;=\; 0 \tag{2.5}$$
$$\text{or } f_3 - f_1 - f_2 \;=\; 0 \tag{2.6}$$

where the half arrows pointing toward the 1-junction are positive forces and those pointing away are negative forces.

Since all velocities are the same as v, we have the power relation of

$$f_1 v + f_2 v - f_3 v \;=\; 0 \tag{2.7}$$
$$\text{or } f_3 v - f_1 v - f_2 v \;=\; 0 \tag{2.8}$$

Equations (2.7) and (2.8) state that the net power supplied adds up to zero. Particularly, the input power supplied is the same as the output power delivered. The half arrows point toward the 1-junction for power supplied and point away for power delivered.

Let's suppose that the cart has a mass of m. Then, Equation (2.6) is changed to

$$f_3 - f_1 - f_2 = m\dot{v} = f_I \tag{2.9}$$

where f_I is the inertial force generated by the mass, which is equal to $\dot{p}$. Equation (2.9) states that the cart can be represented in the bond graph using a 4-port with a 1-junction, as shown in Figure 2.5. A passive 1-port, I is shown together with a 1-junction having the property of the momentum, p, and velocity, v relation, which is called a constitutive relation associated with the parameter m, that is $v = p/m$.

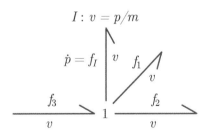

Figure 2.5: A bond graph using a 4-port with 1-junction when the cart has a mass m.

0-junction (common force junction)

Figure 2.6 (a) shows a damper connected between two carts with interacting forces. Figures 2.6(b) or (c) show 3-port 0-junction diagrams to express that the forces of the three points of velocity acting on the damper are the same.

$$f_1 = f_2 = f_3 = f$$

where f is a common force. The velocity relations of the free-body diagram are respectively obtained as

$$v_1 - v_2 - v_3 = 0$$
$$\text{or} \ \ v_2 + v_3 - v_1 = 0$$

The half arrows point toward the 0-junction for power supplied and away for power delivered.

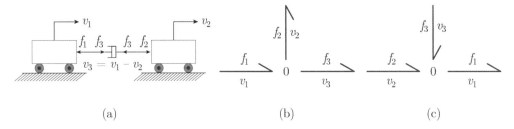

(a)   (b)   (c)

Figure 2.6: (a) A cart on rollers where three velocities are associated and (b) 3-port 0-junction diagram to express the forces of the three points of velocity acting on the damper, and (c) the same 3-port 0-junction diagram.

2.2.3 A bond graph representation from the force diagram

Figure 2.7(a) shows a typical example of a mechanical system composed of a mass, spring, and damper. To figure out how a bond graph for a typical mechanical system is constructed, a force diagram is also considered for comparison purposes, as shown in Figure 2.7(b). We define external force f_1 the same as the summation of the damping force, f_d, and spring force f_s. Then, $f - f_1$ is the same as the inertial force, $m\ddot{x}$ from Newton's law, which is represented as

$$f - f_1 \ = \ m\ddot{x} \tag{2.10}$$
$$f_1 \ = \ f_s + f_d = c\dot{x} + kx \tag{2.11}$$

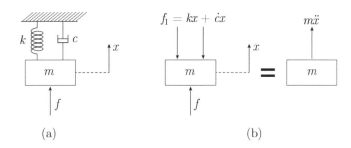

(a)   (b)

Figure 2.7: (a) A typical example of a mechanical system composed of a mass, spring, and damper and (b) a force diagram for a bond graph representation.

Figure 2.8 shows the bond graph, including the information obtained in Equations (2.10) and (2.11). The bonds with half arrows are used to indicate the direction of the power or energy. It shows that input power or energy is transferred as it is stored in the I and C elements and is dissipated in the R elements.

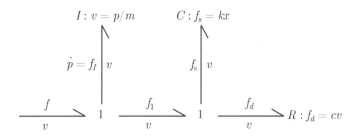

Figure 2.8: Bond graph of a mechanical system composed of mass, spring, and damper.

In more detail, the force f is first applied to the system as an effort source. Next, the inertia force, $f_I = f - f_1$, acts on mass m, which was also previously represented in Newton's law using Equation (2.10). Here, the force relation of $f_I = f - f_1$ is represented using the 1-junction. The energy transmitted to the mass is stored in the passive I element. Then, velocity v of the mass is obtained as a result of momentum p applied to mass m, which is represented by the constitutive relation, $v = p/m$. Here, p is just the integral of the force difference $(f - f_1)$ acting on the mass. Then, the force f_1 with velocity v is delivered to other components, R and C elements. Finally, when v is applied to the damper and spring, f_s and f_d are determined using the constitutive relations, $f_s = kx$ and $f_d = cv$, respectively. Here, the force relation of $f_1 = f_s + f_d$ is represented using the 1-junction, which was also previously represented using

Equation (2.11). Here, 1-junction is used to indicate that the mass, spring, and damper have the same velocity. Force summation of in-going bonds is the same as that of out-going bonds. The 1-junctions connected in a row can be combined and equivalently expressed as shown in Figure 2.9.

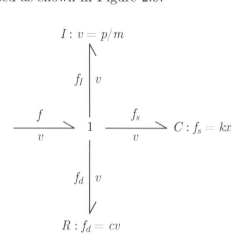

Figure 2.9: Bond graph of a mechanical system shown in Figure 2.8.

Example 1: Figure 2.10(a) shows another mechanical system composed of masses m_1, m_2, springs k_1, k_2, and dampers c_1, c_2 connected in series. Force f is applied to mass m_1. f_{s1} and f_{s2} are the spring forces generated due to k_1 and k_2, respectively. f_{d1} and f_{d2} are the damping forces generated due to c_1 and c_2. x_1 and x_2 are the displacements of m_1 and m_2, respectively. v_1 and v_2 are the velocities of m_1 and m_2, respectively. p_1 and p_2 are the momentum of m_1 and m_2, respectively. Draw its corresponding bond graph.

Solution 1: Figure 2.10(b) shows a bond graph model demonstrating how the force and velocity are delivered through the components using 1-junctions and 0-junctions. Figure 2.11 shows an alternative bond graph of the mechanical system shown in Figure 2.10(a). The difference between the two bond graphs is whether f_{s1} and f_{s2} act on m_1 individually using two 0-junctions or they add up and act on m_1 using the single 0-junction and 1-junction. It is convenient to determine the 1-junctions for the I elements first because the forces acting on masses produce the same velocities at each bond. The 0-junctions are generally associated with the C elements because the velocities at both ends of the spring are different.

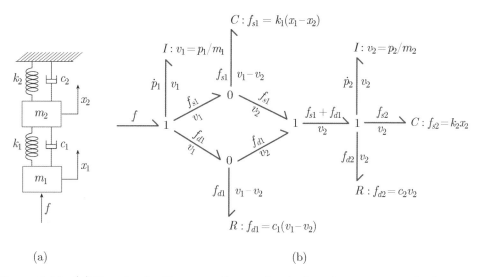

(a)   (b)

Figure 2.10: (a) Free-body diagram of a mechanical system composed of a mass, spring, and damper connected in series and (b) its corresponding bond graph.

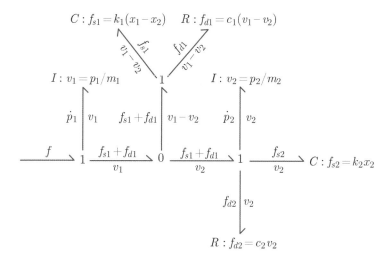

Figure 2.11: An alternative bond graph of the mechanical system shown in Figure 2.10(a).

In summary, the bond graph modeling method has many advantages compared to other conventional methods. First, we can easily recognize which components are used for modeling because they are graphically depicted at ports. Second, the bond graph shows which input type is chosen between the effort and flow. Third, it shows where the power or energy is stored and consumed using the $I, C,$ and R elements. Lastly, the bond graph also shows where the power or energy is delivered through the elements. It indicates graphically that the input power will eventually decrease as it encounters the elements because the mass, spring, and damper act as loads against the applied force or velocity.

Q1: Draw bond graphs of mechanical systems shown in Figures 2.12(a) and 2.12(b).

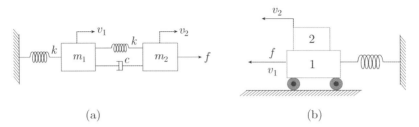

(a)   (b)

Figure 2.12: (a) A first mechanical system and (b) second mechanical system.

Q2: Draw bond graphs of the mechanical systems shown in Figure 2.13(a) and 2.13(b).

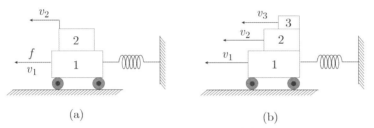

(a)   (b)

Figure 2.13: (a) A first mechanical system and (b) second mechanical system.

Q3: Draw bond graphs of the mechanical systems shown in Figure 2.14(a) and 2.14(b).

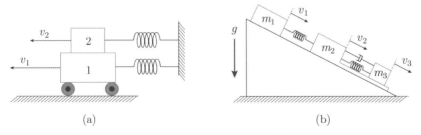

(a)   (b)

Figure 2.14: (a) A first mechanical system and (b) second mechanical system.

Q4: Draw a bond graph of the mechanical systems shown in Figure 1.11.

2.2.4   2-port elements

Transformer

Energy transmission to a mass, spring, and damper is called a 1-port element because there is one port associated with them. The power is reduced through these elements. However, there are elements where the power is transformed into another power without power loss. One of these elements is called a transformer, indicated as T, which is a 2-port element because there are two ports associated with the energy transmission.

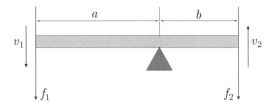

Figure 2.15: A lever system working as a transformer.

For the lever shown in Figure 2.15, the effort (force) at the left hand, f_1 is proportional to the effort (force) at the right hand, f_2, which is obtained from the condition of the same moment about the pivot, assuming that the lever is mass-less [7].

$$af_1 = bf_2 \tag{2.12}$$

Equation (2.12) is rewritten as follows:

$$f_2 = \frac{a}{b}f_1 \tag{2.13}$$

Another relation between v_1 and v_2 at both ends is obtained from the condition of the same angular velocity ω of the lever about the pivot as

$$\frac{v_1}{a} = \frac{v_2}{b} = \omega \tag{2.14}$$

Equation (2.14) is rewritten as follows:

$$v_2 = \frac{b}{a}v_1 \tag{2.15}$$

The ratio of $\frac{a}{b}$ is called a modulus.

The bond graph of the lever is shown in Figure 2.16. The transformer in a mechanical system can be also found in gears, racks and pinions, and pulleys. Using Equations (2.13) and (2.15), we know that the power through the transformer is conserved because of the relation of $f_1 v_1 = f_2 v_2$

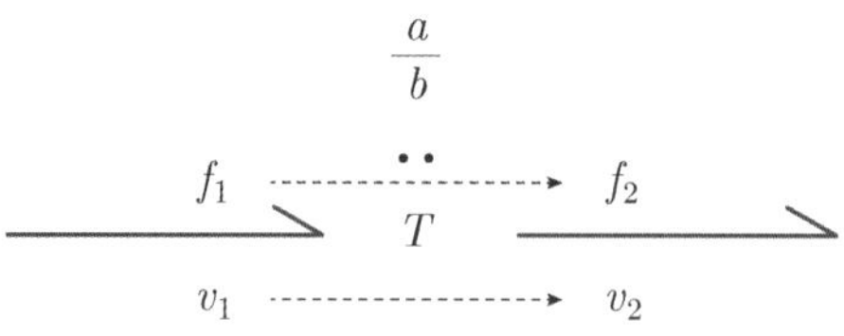

$$\frac{a}{b}$$

$$f_1 \dashrightarrow \overset{\bullet\bullet}{T} \to f_2$$

$$v_1 \dashrightarrow v_2$$

Figure 2.16: A bond graph of the transformer.

Example 2: Figure 2.17 shows a pulley system where m_1 is pulling down due to gravity. The mass m_2 is moving by the wire attaching to the pulley. When the velocities of m_1 and m_2 are v_1 and v_2, what is the velocity of m_2 that is the same as $\frac{dx_2}{dt}$?

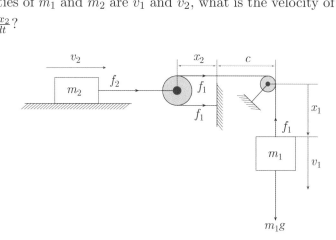

Figure 2.17: A pulley system represented by a transformer.

Solution 2: From Figure 2.17, the force relation is represented as

$$2f_1 = f_2$$

In addition, the displacement relation is also represented as follows, assuming the cable length is constant

$$2x_2 + c + x_1 = \text{constant} \tag{2.16}$$

By differentiating Equation (2.16), we obtain $|v_2| = \frac{1}{2}|v_1|$, The bond graph of the pulley is constructed as shown in Figure 2.18. From the kinematic relations, the transformer has a modulus of two. Here, I_1, I_2, and R represent the 1-port elements associated with mass m_1, mass m_2, and the friction force acting on m_2 developed on the floor surface.

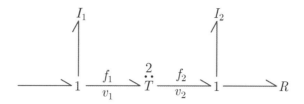

Figure 2.18: A bond graph of the pulley system.

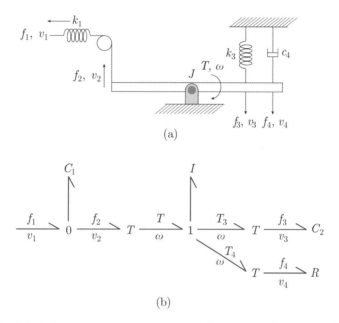

Figure 2.19: (a) A lever system represented by a transformer and (b) its corresponding bond graph.

Figure 2.19(a) shows another lever system with its both ends connected to springs and a damper. The applied input is the velocity of the spring, v_1. When the moment of inertia of the lever J needs to be considered in the modeling, a bond graph model with a transformer can be constructed as shown in Figure 2.19(b). Here, C_1, C_3, I, and R represent the 1-port elements associated with the springs k_1 and k_3, the moment of inertia of the lever J, and damper c_4, respectively. The T-port is used for linear motion to rotational motion and vice versa.

Q5: Draw a bond graph for a system composed of the wheels, a rack and pinion, and a spring when torque T_1 is applied to one of the wheels as shown in Figure 2.20 where the graphical definitions are denoted.

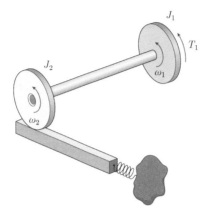

Figure 2.20: A mechanical system composed of wheels, a rack and pinion, and a spring.

Gyrator

There is another 2-port system, a gyrator whose name is derived from the gyrational coupling terms that frequently occur in nonlinear mechanical systems [3]. In the transformer, the effort variables at the two ports are proportional to each other and the flow variables are also proportional to each other. For a gyrator indicated by G, the effort of one port is proportional to the flow of the other port and vice versa. Particularly, if the modulus is r, we have the relations of

$$f_1 = rv_2 \tag{2.17}$$
$$rv_1 = f_2 \tag{2.18}$$

As in the case of the transformer, the power is also conserved. This can be seen by multiplying the left and right sides of Equations (2.18) and (2.18),

$$f_1 v_1 = f_2 v_2$$

The gyrator bond graph is shown in Figure 2.21. The cross-relation is indicated using the dashed arrow for the gyrator, which is indicated by G with a radius r

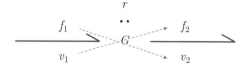

Figure 2.21: A bond graph of a gyrator.

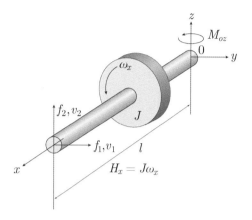

Figure 2.22: A gyroscope.

A gyrator example can be found in a mechanical gyroscope shown in Figure 2.22 [3]. Suppose that a wheel having inertia J is rotating around the x axis with angular velocity ω_x. Then, it has an angular momentum $H_x = J\omega_x$. When force f_1 causes a moment $M_{oz} = f_1 l$ around z axis, We know that the moment about the fixed point must equal the time rate of angular momentum around O, $\dot{H}_o$. Thus, we have

$$f_1 l = M_{oz} = |\dot{\mathbf{H}}_0|$$

Hence, $\dot{\mathbf{H}}_0$ must also be in the z-axis direction and have the magnitude of H_x. This is accomplished by having H_0 swivel so that the rod tip moves upward with velocity v_2. In a component form, the relation is

$$f_1 l = M_{oz} = J\omega_x \frac{v_2}{l}$$

or

$$f_1 = \frac{J\omega_x}{l^2} v_2 \tag{2.19}$$

Equation (2.19) states that the force of port 1 is proportional to the velocity of port 2 having a modulus of $\frac{J\omega_x}{l^2}$.

A similar analysis for the effect of f_2 yields

$$f_2 = \frac{J\omega_x}{l^2}v_1 \qquad (2.20)$$

Equation (2.20) indicates that the force of port 2 is proportional to the velocity of port 1 and has the same modulus. The gyrator will be studied again in Chapter 5 for power conversion in electromechanical systems.

2.3 Causality

We begin this section by first asking whether a mechanical variable, such as force and velocity can be simultaneously applied to the mechanical components such as a mass or spring. Further, a question can also be raised as to whether electrical variables, such as voltage and current, can be simultaneously applied to electrical components, such as a resistor or capacitor. Particularly, is it possible to specify the velocity of a mass if a known force is applied to it? Or is it possible to specify the force applied to a spring if a known velocity is applied to it? For example, when $1\,N$ force is applied to a $1\,kg$ mass, as shown in Figure 2.23(a), is it possible to simultaneously specify $10\,m/sec$ velocity to the mass? The answer is definitely no. For this reason, the velocity is consequently determined by integrating the acceleration, which is obtained using the applied force divided by the mass with respect to time. This fact is also found in an electrical circuit, as shown in Figure 2.23(b). It is impossible to apply $1\,V$ and $1\,A$ simultaneously to the $10\,\Omega$ resistor because the current is determined as a result of applying $1\,V$ to the resistor.

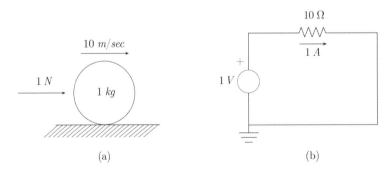

(a)   (b)

Figure 2.23: Causality of sources for (a) mechanical and (b) electrical systems.

The interaction between the effort and flow is reflected in a bond graph using causality. Causality is the determination of input and output variables of elements on a bond graph. The primary restriction of causality can be stated as "either the effort or flow must be for a port but both cannot be input or output simultaneously on the same port [3]." Particularly, it is impossible to simultaneously specify the force and velocity.

If the effort is input from a component A to a component B, then the flow is obtained as a result. Then, the flow is input from a component B to a component A and vice versa. The effort application is indicated using a bar attached to a bond. The flow application is indicated using a point on the bond. The bar and point on the bond are graphically understood by visualizing "effort pushes and flow points" [3], as shown in Figure 2.24.

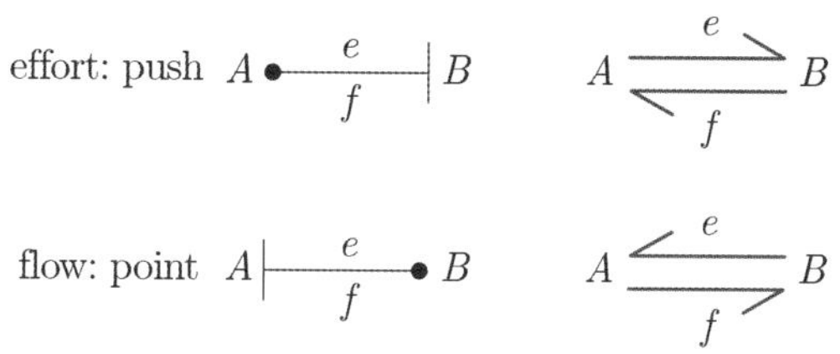

Figure 2.24: Causality marks of effort and flow using a bar and a point.

2.3.1 Source causality

By the definition of the effort and flow, the proper causalities of the effort and flow sources are represented with the half arrow bond for power direction appeared together as shown in Figure 2.25.

Figure 2.25: Indication of effort and flow by using a bar and a point.

2.3.2 Causalities of energy storage elements

For energy source elements, namely C and I-port elements, there are two types of causalities. Consider the C element first. If f (flow) is the input to a C element, it is integrated to provide q (displacement), and then, e is obtained in return as a function of q ($e = \phi_s(q)$). Hence, this C-port is called an integral causality. This situation in graph and equation form is shown in Figure 2.26(a). However, if e is the input to a C element, q is obtained by a function of ϕ_s^{-1}, then f is obtained in return by differentiating q. That is why the C-port in this case is called a derivative causality. This situation in graph and equation form is shown in Figure 2.26(b). The same concept applies to the I element. If e is

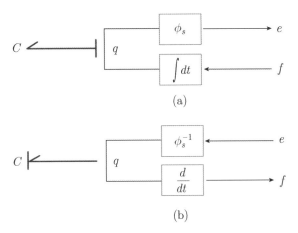

Figure 2.26: (a) Integral and (b) derivative causality C-port.

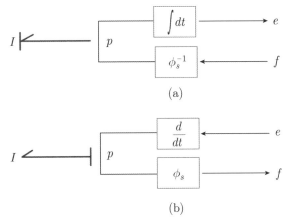

Figure 2.27: (a) Integral and (b) derivative causality I-ports.

the input into an I element, p is obtained by integrating e, thus, f is obtained in return as a function of p ($f = \phi_s(p)$). Hence, this I port is called an integral causality I port. This situation in graph and equation form is shown in Figure 2.27(a). However, if f is the input into an I element, p is obtained as a function of ϕ_s^{-1}, thus, e is obtained in return by differentiating p. That is why the I-port in this case is called a derivative causality. This situation in graph and equation form is shown in Figure 2.27(b).

The derivative action causes a problem when an abrupt input is applied as it is easily understood from its mathematical description. When there are derivative causalities in mechanical structures, they experience a large impact, big oscillation, and saturation, which eventually causes their breakdown. For example, if a step flow input is applied to an I element, infinity force due to the derivative causality can be produced. Hence, we should take caution when designing mechanical systems so that they have no derivative causalities in I and C elements.

2.3.3 Causality of energy dissipating element

Causality for R can be imposed in either direction as shown in Figures 2.28(a) and (b). When effort e is applied to R, flow f is determined in return. When flow f is applied to R, effort e is determined in return.

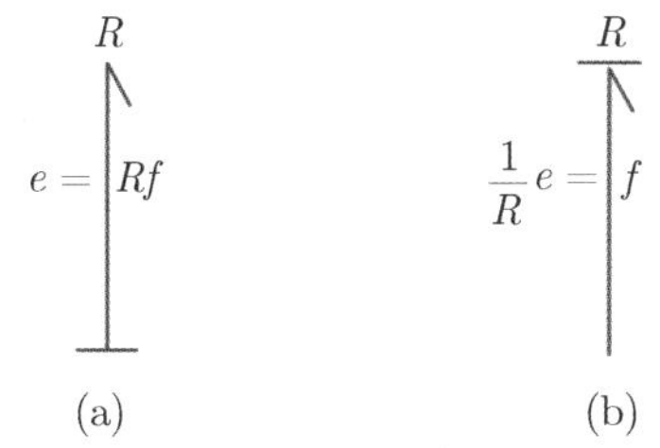

Figure 2.28: Causality for R imposed in either direction.

2.3.4 Causality of junction elements

The 0-junction has three possible causal patterns when there are 3-ports. We consider one of the three cases where two flow inputs, f_1 and f_3 are given as shown in Figure 2.29(a). We have the following relations associated with the 0-junction

$$e_1 = e_2 = e_3 \tag{2.21}$$

$$f_1 - f_3 = f_2 \tag{2.22}$$

Thus, as two flow inputs, f_1 and f_3 are given, the third flow f_2 is determined and it is an input to the element. As a result, the effort should be obtained. Hence, the causality should be the one shown in Figure 2.29(b).

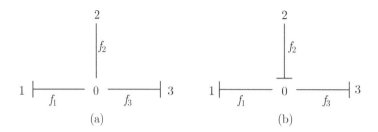

Figure 2.29: Causality of the 0-junction (a) when flow input f_1 and (b) f_3 are given.

In the second case, suppose that f_1 and f_2 are impressed on the 0 junction as inputs, then, the third flow must be output because of the flow relation. Hence, the third effort must be determined in return, as shown in Figure 2.30. In conclusion, there should be only one bar at the bonds directing to the 0 junction. The resultant force is then obtained as

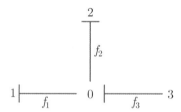

Figure 2.30: Causality of the 0-junction how f_3 causality is determined when f_1 and f_2 are given.

Q6: What is the causality bond when e_1 is applied to 0 junction having three ports?

As an example of causality assignment of the 0-junction, we can think of a spring to which velocity inputs, f_1 and f_3 are given, as shown in Figure 2.31(a). The flow difference, $f_2 = f_3 - f_1$ is an input to the C element and indicated using the point mark. Then, spring force e_2 is obtained as a return and indicated using the bar mark, as shown in Figure 2.31(b).

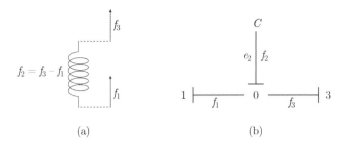

(a)   (b)

Figure 2.31: (a) A spring to which two flow inputs, f_1 and f_3 are applied and (b) causality assignment of the 0-junction.

1-junction causal patterns are found using the same reasoning, with efforts and flows interchanged. As an example, suppose that efforts e_1 and e_3 are applied at the 1-junction, as shown in Figure 2.32(a). We have the following

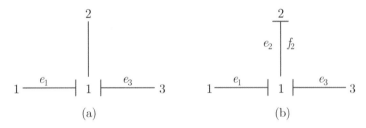

(a)   (b)

Figure 2.32: Causality of the 1-junction when e_1 and e_3 are given.

relations associated with the 1-junction:

$$f_1 = f_2 = f_3 \tag{2.23}$$

$$e_1 - e_3 = e_2 \tag{2.24}$$

Thus, as two effort inputs, e_1 and e_3 are given, the third effort e_2 is determined and it is an input to the element. As a result, flow f_2 is determined. Hence, the causality should be the one shown in Figure 2.32(b).

As an example of causality assignment of the 1-junction, we can consider a mass to which two force inputs e_1 and e_3 are given, as shown in Figure 2.33(a). The relative force difference $(e_1 - e_3)$ is an input to the I element and indicated using the bar mark. Then, velocity f_2 is obtained as a return and indicated using the point mark as shown in Figure 2.33(b). In conclusion, there should be only one point at the bonds directed toward the 1 junction.

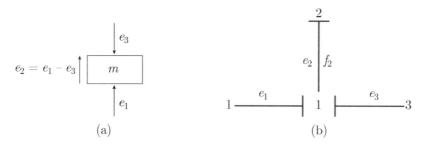

(a) (b)

Figure 2.33: (a) A mass to which two effort inputs, e_1 and e_3 are applied and (b) causality assignment of the 1-junction.

2.3.5 Causality of 2-port elements

When effort e_1 on one bond is given, effort e_2 on the remaining bond is specified in the T element, while flow f_1 on one bond is given, a flow f_2 on the remaining bond is specified. Therefore, the causality should be the one shown in Figures 2.34(a) and (b).

(a) (b)

Figure 2.34: Causality of T element (a) when effort e_1 is given, effort e_2 is specified and (b) when flow f_1 is given, flow f_2 is specified.

When effort e_1 on one bond is given, flow f_2 on the remaining bond is specified in the G element while flow f_1 on one bond is given, effort e_2 on the remaining bond is specified. Therefore, the causality should be the one shown in Figures 2.35(a) and (b).

(a) (b)

Figure 2.35: Causality of the G element (a) when effort e_1 is given, flow f_2 is specified and (b) when flow f_1 is given, effort e_2 is specified.

The causality assignment procedure is summarized by referring to [3] as follows:

1. Choose S_e or S_f and assign it required causality. Immediately extend the causal implications of this action, using 0- and 1-junction restrictions.

2. Repeat step 1 until all sources have been causally assigned.

3. Choose any C or I and assign integral causality to it. Again, extend the causal implications of this action, using 0- and 1-junction restrictions.

4. Repeat step 3 until all C or I elements have been causally assigned.

5. Choose any R that is unassigned and give it an arbitrary causality. Extend the causal implications of this action using 0- and 1-junction restrictions.

6. Repeat step 5 until all R elements have been causally assigned.

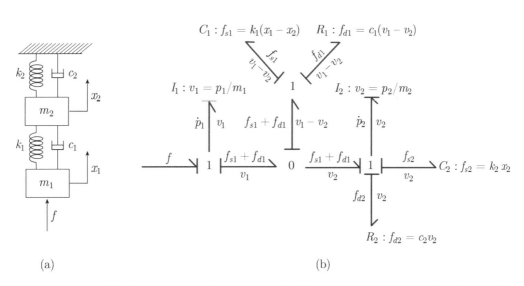

Figure 2.36: (a) A mechanical system and (b) its corresponding bond graph with causality marks added.

Based on the causality rules, a mechanical system shown in Figure 2.36(a) is drawn with causality marks added to the bond graph in Figure 2.36(b). Using the causality marks, we can see how velocity and force are respectively obtained in the springs, dampers, and masses of the mechanical system. First, the force input, f, acting on mass m_1 is represented as an effort source using the causality mark in the bond. Second, a 1-junction is used to represent the force difference, $f - (f_{s1} + f_{d1})$, which is the same as $\dot{p}_1$ applied to the I_1 element. Then, velocity v_1 is obtained as a result using the constitutive relation of $v_1 = \frac{p_1}{m_1}$. The determined velocity, v_1, is delivered to the next C_1 and R_1 elements. Third, a 0-junction is used to indicate that force $f_{s1} + f_{d1}$ generated at one end of k_1 and c_1 is the same as that on the other end. Fourth, the displacement difference $(x_1 - x_2)$ and velocity difference $(v_1 - v_2)$ separately apply to the C_1 and R_1 elements, respectively. As a result, f_{s1} and f_{d1} are obtained using constitutive relations of $f_{s1} = k_1(x_1 - x_2)$ and $f_{d1} = c_1(v_1 - v_2)$, respectively. Fifth, force $f_{s1} + f_{d1}$ is delivered to the next I_2 element. A 1-junction is used to indicate that the force difference, $(f_{s1} + f_{d1}) - (f_{s2} + f_{d2})$, is the same as $\dot{p}_2$ applied to another I_2 element. Then, velocity v_2 is obtained as a result using the constitutive relation of $v_2 = \frac{p_2}{m_2}$. Finally, the determined velocity v_2 is delivered to the C_2 and R_2 elements. As a result, $f_{s2} = k_2 x_2$ and $f_{d2} = c_2 v_2$ are respectively obtained using the corresponding constitutive relations.

It is observed that all I and C elements have integral causalities in the bond graph where forces apply to masses and velocities are obtained as a result, whereas velocities apply to springs and forces are obtained as a result. Hence, this mechanical system behaves properly.

Q7: Draw bond graphs with causality marks assigned for Figures 2.12, 2.13, and 2.14.

2.4 State equations

2.4.1 When integral causality elements only exist

A modeling process is complete when we can obtain a set of describing equations (usually differential equations). One remarkable feature of the bond graph is that the equation formulation process can be directly obtained from the bond graph itself where causality and constitutive relations are represented. In conventional approaches, we need to know how to use Newton's law to derive the governing equations.

Bond graphs lend themselves to state space methods, in which an nth-order dynamic system is represented by n first-order differential equations in n variables [3]. The state equations are closely connected to energy storage elements I and C in the system. The variables in bond-graph equation sets will always be energy variables p and q, and the state equations are effort $\dot{p}$ and flow $\dot{q}$

equations. Hence, the number of state equations is the same as the number of integral causalities in the bond graph.

For general nonlinear models, we write equations as follows:

$$\dot{X}_1 = \phi_1(X_1, X_2, \cdots X_n; u_1, u_2, \cdots u_r)$$
$$\dot{X}_2 = \phi_2(X_1, X_2, \cdots X_n; u_1, u_2, \cdots u_r)$$
$$\cdots\cdots\cdots\cdots\cdots\cdots\cdots\cdots\cdots\cdots\cdots\cdots$$
$$\dot{X}_n = \phi_n(X_1, X_2, \cdots X_n; u_1, u_2, \cdots u_r)$$

where

X_i = state variables (energy variables)

$\dot{X}_i$ = derivatives of state variables (efforts or flows) with respect to time

u_j = effort or flow sources

ϕ_i = algebraic functions

For a simpler appearance, we use a matrix form

$$\dot{X} = A\mathbf{X} + B\mathbf{u} \tag{2.25}$$

where $\mathbf{X}$ and $\mathbf{u}$ are column vectors. A and B are the system matrices.

Let us represent the state equations for a system shown in Figure 2.37(a) whose corresponding bond graph is shown in Figure 2.37(b). The bond graph is displayed again with the state variables, i.e., energy variables, $X_1 = x - x_1$, p_1, $X_2 = x_1 - x_2$, and p_2. As all energy-storing elements (all 2 C's and 2 I elements) are integral causality, there are four state equations having four state variables, X_1, P_1, X_2, and P_2.

$$\dot{X}_1 = \dot{x} - \dot{x}_1 : f_1 = k_1 X_1 \tag{2.26}$$
$$\dot{p}_1 = f_1 - f_2 : \dot{x}_1 = \frac{p_1}{m_1} \tag{2.27}$$
$$\dot{X}_2 = \dot{x}_1 - \dot{x}_2 : f_3 = k_2 X_2, f_2 = f_3 + f_4, f_4 = c_2 \dot{X}_2 \tag{2.28}$$
$$\dot{p}_2 = f_2 : \dot{x}_2 = \frac{p_2}{m_2} \tag{2.29}$$

Rearranging through Equations (2.26) and (2.29), we obtain the state equations represented by all energy variables as

$$\dot{X}_1 = \dot{x} - \frac{p_1}{m_1}$$

$$\dot{p}_1 = k_1 X_1 - k_2 X_2 - c_2 \dot{X}_2 = k_1 X_1 - k_2 X_2 - c_2 \left(\frac{p_1}{m_1} - \frac{p_2}{m_2} \right)$$

$$\dot{X}_2 = \frac{p_1}{m_1} - \frac{p_2}{m_2}$$

$$\dot{p}_2 = k_2 X_2 + c_2 \left(\frac{p_1}{m_1} - \frac{p_2}{m_2} \right)$$

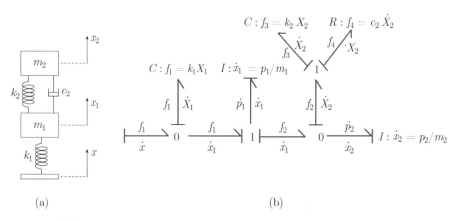

(a) (b)

Figure 2.37: (a) A system having two masses and two springs and (b) its corresponding bond graph.

For comparison purposes, the governing equations are also obtained by applying Newton's law to masses m_1 and m_2 as follows:

$$m_1 \ddot{x}_1 = k_1(x - x_1) - k_2(x_1 - x_2) - R(\dot{x}_1 - \dot{x}_2)$$

$$m_2 \ddot{x}_2 = k_2(x_1 - x_2) + R(\dot{x}_1 - \dot{x}_2)$$

From the definition of $\dot{p}_1 = m_1 \ddot{x}_1$ and $\dot{p}_2 = m_2 \ddot{x}_2$, the state equations are exactly the same equations as those obtained by applying Newton's method. Furthermore, the n first-order equations are generally simpler in form than the higher-order equations derived from Newton's law.

Q8: Obtain the state equations for the systems shown Figure 2.12, 2.13, and 2.14.

2.4.2 When derivative causality elements exist

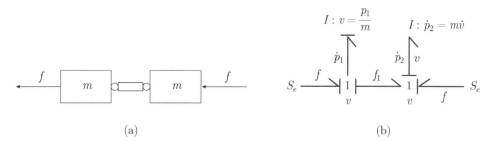

(a)   (b)

Figure 2.38: (a) A schematic diagram and (b) a bond graph of trains having a derivative causality.

Let's now consider how the state equation is obtained when derivative causality exists in the bond graph. For example, there is a train loosely connected with a rigid link shown in Figure 2.38 (a). From its bond graph shown in Figure 2.38 (b), we know there is derivative causality. As there is one integral causality, there must be one state equation, which is derived using the constitutive relation as

$$\dot{p}_1 = f - f_1; \ v = \frac{p_1}{m} \tag{2.30}$$

However, we also have the following relation from another I element:

$$\dot{p}_2 = f + f_1; \ \dot{p}_2 = m\dot{v} \tag{2.31}$$

Using Equation (2.30) and Equation (2.31), the state equation is obtained as

$$\dot{p}_1 = f - (\dot{p}_2 - f) = 2f - \dot{p}_2 = 2f - m\dot{v} = 2f - \dot{p}_1 \tag{2.32}$$

Hence, we have $\dot{p}_1 = f$. The result is the same as the result obtained by applying $2f$ to two connected masses having mass, $2m$.

If the velocity, v, is abruptly changed (for example, when the rod and masses are not tightly connected), the relation $\dot{p}_2 = m\dot{v}$ expressed in Equation (2.31) can not be defined because of non-differentiable velocity $\dot{v}$. As a result, Equation (2.32) is no longer valid, which indicates that the state equation does not describe the correct system behavior.

As shown in the above example, we can obtain the state equation regardless of the existence of the derivative causality. However, if a system or plant has a derivative causality element, it will cause unstable operation. Bond graph modeling provides the advantage of knowing whether or not a system has a causality problem.

Q9: Draw a bond graph of the trains connected by rigid links shown in Figure 2.39 when a step force is applied to the first train. What will happen to trains? How can we fix it to avoid mechanical problems?

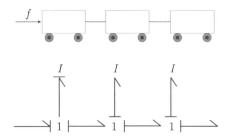

Figure 2.39: A schematic diagram (top) and bond graph (bottom) of a train connected by rigid links.

Figure 2.40(a) shows a more complicated mechanical structure composed of masses, spring, damper, and lever with velocity and force inputs, v and f, respectively. Figure 2.40(b) shows its corresponding bond graph. Let us determine whether it has derivative causalities and its state equations.

The bond graph is displayed using the numbers denoted on the bonds for easy differentiation. Using these numbers, let us suppose the power variables (effort and flow) of each bond are denoted as (e_1, f_1), (e_2, f_2), $\cdots \cdots$, (e_9, f_9). Similarly, the energy variables (momentum and displacement) of each bond are denoted as (p_1, q_1), (p_2, q_2), $\cdots \cdots$, (p_9, q_9). From the causality representation, we know that there are two integral causalities, C_2 and I_8, one derivative causality, I_4. Thus, we obtain two state equations using the numbers assigned to the elements, as follows:

$$\dot{q}_2 = f_1 - f_3 : e_2 = \frac{1}{C_2} q_2 \tag{2.33}$$

$$\dot{p}_8 = e_9 + e_6 - e_7 : f_8 = \frac{1}{I_8} p_8 \tag{2.34}$$

There are many relations obtained by 0- and 1-junctions. Moreover, the transformer having modus m has the following relations:

$$
\begin{aligned}
e_2 &= e_3 = e_1 \\
e_3 &= e_4 + e_5 \\
f_3 &= f_4 = f_5 = mf_6 = mf_7 = mf_8 \\
e_6 &= me_5
\end{aligned}
\tag{2.35}
$$

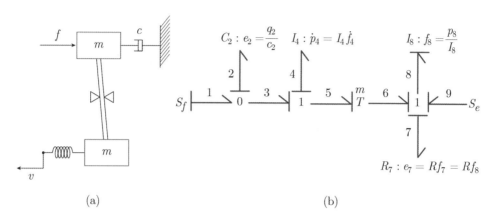

(a)   (b)

Figure 2.40: (a) A schematic diagram and (b) a bond graph of a mechanical structure composed of masses, a spring, and a lever.

For the derivative causality I_4 element, in which f_4 is applied to I_4, e_4 is thus determined using the following relation:

$$
e_4 = \dot{p}_4 = I_4 \dot{f}_4
\tag{2.36}
$$

Using the above relations, we can obtain the state equation for $\dot{p}_8$. To explain how it is obtained, the process is described in more detail below.

$$
\begin{aligned}
\dot{p}_8 &= e_9 + m(e_2 - e_4) - e_7 \tag{2.37} \\
&= e_9 + m\left(\frac{1}{C_2}q_2 - \dot{p}_4\right) - R_7 f_8 \tag{2.38} \\
&= e_9 + m\left(\frac{1}{C_2}q_2 - I_4 \dot{f}_4\right) - R_7 f_8 \tag{2.39} \\
&= e_9 + m\left(\frac{1}{C_2}q_2 - mI_4 \dot{f}_8\right) - R_7 f_8 \tag{2.40} \\
&= e_9 + m\left(\frac{1}{C_2}q_2 - m\frac{I_4}{I_8}\dot{p}_8\right) - R_7 f_8 \tag{2.41}
\end{aligned}
$$

Thus, we finally obtain the state equation for q_2 and p_8 as follows:

$$\dot{q}_2 = f_1 - m\frac{p_8}{m_8} \qquad (2.42)$$

$$\dot{p}_8 = \frac{mC_2^{-1}}{1 + m^2 I_4 I_8^{-1}}q_2 - \frac{R_7 I_8^{-1}}{1 + m^2 I_4 I_8^{-1}}p_8 + \frac{1}{1 + m^2 I_4 I_8^{-1}}e_9 \qquad (2.43)$$

From the above examples, we know that the state equations can be concurrently obtained in a different way from that of the integral causality case. We can obtain the state equations even though it looks complicated. However, if velocity f_4 is abruptly changed, the relation expressed in Equation (2.36) cannot be defined because of the non-differentiable f_4. As a result, Equations (2.42) and (2.43) are not valid any longer, which indicates that the state equation does not describe the correct system behavior.

Q10: How can we change the model of the mechanical system shown in Figure 2.40 to avoid its derivative causality problem?

Bond graph modeling for electrical systems

—

Bond graph modeling for electrical systems

3.1   Constitutive relations

It was stated in Chapter 1 that there are only three parameters in electrical systems: inductor L, resistance R, and capacitor C. Moreover, there are four variables: voltage e, current i, flux linkage λ, and charge q, which are equivalent to force f, velocity v, momentum p, and displacement x in a mechanical system. e and i are called power variables because the product of e and i is a power. Additionally, the power and energy variables of various physical systems are listed in Table 1.1.

3.2   Bond graph model of electric circuits

For representing the constitutive relations in a more general form, which shows similarities with other physical systems, a bond graph for a passive 1-port is introduced in Figure 3.1. The graphical notations of the capacitor, resistor, and inductor are denoted using C, R, and L, respectively. The power variables, e and i are denoted at the left and right parts of the bond. The half arrow notations are used to express the power flow directions and the bar notation is used to include the causality relations among the four variables. For example, in the C port, when the current passing through capacitor i_c is applied, capacitor voltage e_c is generated using the corresponding constitutive relation in return. In the R port, either current or voltage can be applied, and voltage or current is generated in return. In the L port, when voltage e_L is applied, current i_L is generated using the corresponding constitutive relation in return.

$C\!:\, e_c = \dfrac{1}{C} q = \dfrac{1}{C} \int i_c\, dt$	$R\!:\, e_R = R i_R \text{ or } i_R = \dfrac{e_R}{R}$		$L\!:\, i_L = \dfrac{\lambda}{L} = \dfrac{\int e_L dt}{L}$
C $e_c \mid i_c$	R $e_R \mid i_R$	or $\quad R$ $e_R \mid i_R$	L $e_L \mid i_L$
Current input to a capacitor, then voltage output in return	Current or voltage input to a resistor, then voltage or current output in return respectively		Voltage input to an inductor, then current output in return

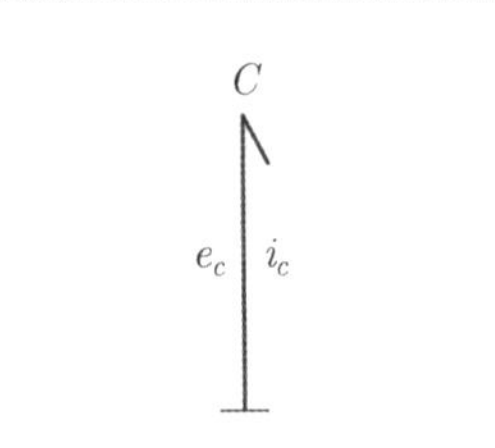

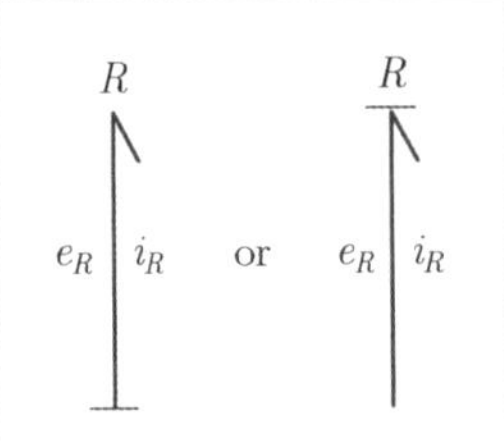

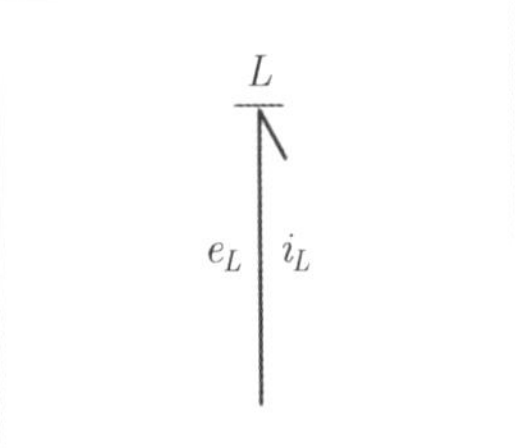

Figure 3.1: Bond graph representation of electrical components.

For reference, electrical energy E is defined by integrating electrical power P with respect to time. Hence, E is represented using the following constitutive relations:

$$E \;=\; \int P dt = \int e i dt = \int e \frac{dq}{dt} dt = \int e\, dq = \frac{1}{2C} q^2 \tag{3.1}$$

$$E \;=\; \int P dt = \int e i dt = \int i \frac{d\lambda}{dt} dt = \int i\, d\lambda = \frac{\lambda^2}{2L} = \frac{1}{2} L i^2 \tag{3.2}$$

As P is a product of e and i, they are called power variables. Similarly, as momentum λ and displacement q appear in the energy relations as indicated in Equations (3.1) and (3.2), they are called energy variables.

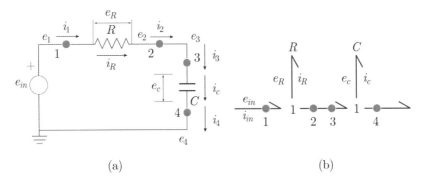

(a) (b)

Figure 3.2: (a) An electrical circuit composed of a power supply, a resistor R, and a capacitor C and (b) its corresponding bond graph representation.

Figure 3.2(a) shows a simple electrical circuit composed of a power supply, a resistor R, and a capacitor C. Circuit theory can be applied to analyze the circuit by applying Kirchhoff's law [8] as follows:

$$e_{in} = Ri_R + \frac{1}{C}\int i_C dt \qquad (3.3)$$

Here, e_{in} is the applied voltage from the power supply. i_1, i_2, i_3, and i_4 are the currents flowing through the points 1, 2, 3, and 4, respectively. i_R, and i_C are the currents flowing through R and C, respectively. e_1, e_2, e_3, and e_4 are the voltages the points 1, 2, 3, and 4, respectively. e_R and e_C are the voltages across R and C, respectively.

Recalling the rules of bond graph representation, the electrical circuit is represented using the bond graph shown in Figure 3.2(b). If the voltage at points 1,2,3, and 4 are denoted as e_1, e_2, e_3, and e_4, $e_1 = e_{in}$ since there is no voltage drop in the conductor. Likewise, $e_2 = e_3$, $e_4 = 0$. If the current at points 1,2,3, and 4 are denoted as i_1, i_2, i_3, and i_4, we know that $i_{in} = i_1 = i_R = i_2 = i_3 = i_C = i_4$. A 1-junction is used to show this condition. We also know that $e_{in} = e_R + e_2$ is based on Kirchhoff's voltage law. This relation is also represented using a 1-junction and the half arrow directions of the in-going and out-going bonds. We also have $e_2 = e_3 = e_C$ because $e_4 = 0$. Thus, we have the following two conditions:

$$i_{in} = i_R = i_C = i \qquad (3.4)$$
$$e_{in} = e_R + e_C = R_1 i + \frac{1}{C}\int i dt \qquad (3.5)$$

From the above relations, the bond graph shown in Figure 3.2(b) is equivalent to the bond graph shown in Figure 3.3.

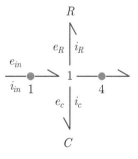

Figure 3.3: An equivalent bond graph shown in Figure 3.2(b).

Figure 3.4 shows another circuit having a parallel connection with the previous circuit. Figure 3.5 shows its bond graph representation. The bond graph shows how the power or energy associated with parameters L, R, and C are delivered in the system. In the electrical circuit, we know that $i_2 = i_C + i_3$ from Kirchhoff's current law, i.e., the current summation at the junction equals zero. A 0-junction is used to show this condition. Additionally, the 0-junction is used to enforce that the voltages around the 0-junction are the same. In summary, it can be said that Kirchhoff's voltage and current laws are implemented using 1- and 0-junctions, respectively.

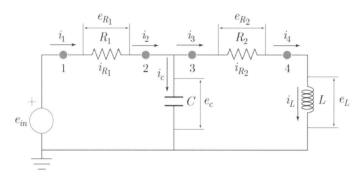

Figure 3.4: Example of an electronic circuit.

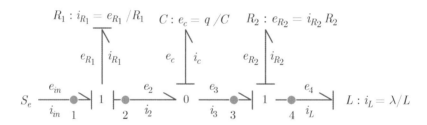

Figure 3.5: Bond graph of the electronic circuit shown in Figure 3.4.

Using the causality marks, we can see how current and voltage are respectively obtained in the resistors, capacitors, and inductors of the electronic circuit. First, the voltage input e_{in} is applied to the electric circuit and is represented as an effort source in the bond. Second, the net voltage $(e_{in} - e_2)$ acting on the resistor R_1 is represented in the bond using 1-port. Then, current i_{R_1} is obtained as a result of the net voltage e_{R_1} applied to the resistor R_1. Third, voltage e_2 with current i_2 is delivered to the next C port. Fourth, i_2 is divided into two different paths, providing i_C and i_3. i_C is applied to the C port and as a result generates voltage e_C which is obtained using the constitutive relation $\frac{q}{C}$. Finally, e_3 is applied to the next elements R_2 and L. Voltage e_3 is used to supply e_4 acting on the L element. Then, i_L is obtained as a result by using the constitutive relations associated with L. Then, the current i_L, which is the same as i_{R_2}, is applied to the R_2 element, which determines e_{R_2} as a result. e_{R_2} is determined using the constitutive relation associated with the resistor R_2. One thing to note is that the inductor, resistor, and capacitor act as loads to the applied voltage or power. Hence, voltage or power will eventually decrease as they encounter these types of loads.

In conclusion, from the 1- and 0-junction characteristics, we obtain the following conditions:

$$i_{in} = i_1 = i_{R_1} = i_2 \tag{3.6}$$
$$e_{in} = e_{R_1} + e_2 \tag{3.7}$$
$$e_2 = e_C = e_3 \tag{3.8}$$
$$i_2 = i_C + i_3 \tag{3.9}$$
$$i_3 = i_{R_2} = i_L \tag{3.10}$$
$$e_3 = e_{R_2} + e_4 \tag{3.11}$$

From the constitutive relations, we obtain

$$i_{R_1} = \frac{e_{R_1}}{R_1} \tag{3.12}$$
$$e_C = \frac{1}{C}q_c = \frac{1}{C}\int i_C dt \tag{3.13}$$
$$e_{R_2} = i_{R_2}R_2 \tag{3.14}$$
$$i_L = \frac{\lambda}{L} = \frac{1}{L}\int e_L dt \tag{3.15}$$

By solving the above differential equations mathematically in a time domain, we determine how the variable of interest in the electrical circuit behaves when the input is applied. The load effect can be easily understood by the bond graph. From the bond graph represented in Figure 3.5, we know that the loads reduce the applied voltage as it encounters several loads, such as R_1, C, and R_2. Hence, e_2 is smaller than e_{in} by the amount of e_{R_1}. Current i_3 is also reduced from i_2 by the amount of i_C. The voltage and current reduced at each load can be obtained using the constitutive relations associated with loads, R_1 and C.

Q1: What are the voltages at points 1, 2, 3, and 4 in Figure 3.4?

3.3 Voltage source and current source in electric circuits

3.3.1 Ideal voltage source

Figure 3.6 shows a simple electrical circuit composed of a voltage source, internal resistor R_{in}, and the load resistor, R_L, connected in series. V_{in} and V_L are defined as voltage drops due to R_{in} and R_L respectively. When an ideal voltage source is defined as a voltage input that provides an input voltage to a load without voltage loss, R_{in} should be small enough to be ignored, i.e., $V_{in} = 0$. However, when R_L is as small as R_{in}, it is impossible to have $V_L = V$. Hence, it is possible to realize an ideal voltage source only when R_L is much greater than R_{in}. A typical example of an ideal voltage source is a function generator generally used for applying various voltage signals to electrical loads. The magnitude of R_L should be checked if it is larger compared to R_{in} to assume $V_{in} \approx 0$.

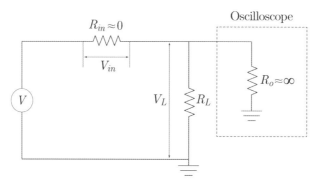

Figure 3.6: A circuit diagram illustrating an ideal voltage source.

For reference, V_L can be measured using an oscilloscope that is modeled as R_o. However, for R_o to have no effect on V_L, R_L must be much smaller than R_o. Otherwise, a correct voltage across R_L is not measured because of the current flowing to R_o. As a result, V_L can be reduced by a voltage division relation involving R_L and R_o. Hence, commercial oscilloscopes typically use $1 \sim 2\, MHz$ resistors.

3.3.2 Ideal current source

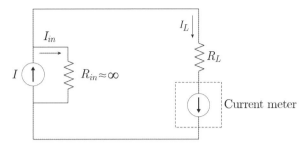

Figure 3.7: A circuit diagram illustrating an ideal current source.

Figure 3.7 shows a simple electrical circuit composed of a current source, internal resistor, R_{in}, and the load resistor, R_L, connected in parallel. I_{in} and I_L are defined as the current flowing through R_{in} and R_L, respectively. When an ideal current source is defined as a current input that provides input current to a load without current loss, R_{in} should be large enough not to flow current through R_{in}, i.e., $I_{in} = 0$. However, when R_L is as large as R_{in}, it is not possible to have $I_L = I$. Hence, it is possible to realize an ideal current source only when $R_L \ll R_{in}$. Here, when I_L needs to be measured using a current meter, it should have a very small impedance to not affect I_L.

3.4 State equations of electrical circuits

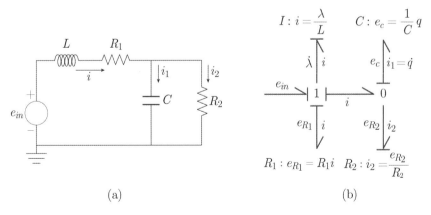

(a)   (b)

Figure 3.8: (a) An electric circuit composed of L, R, and C and (b) its bond graph.

As in the mechanical system, the state variables of electrical systems are energy variables, λ and q. The number of state equations is the same as the number of integral causalities in the bond graph.

For the electric circuit shown in Figure 3.8(a), its corresponding bond graph is shown in Figure 3.8(b). Here, $\dot{\lambda}$, e_{R_1}, and e_{R_2} are the voltage across L, R_1, and R_2, respectively. i, $\dot{q}$, and i_2 are the current flowing through L, C, and R_2, respectively. Since we have two energy-storing elements, I and C with integral causalities, as shown in Figure 3.8(b), there are two state equations having state variables, λ and q as

$$\dot{\lambda} = E_{in} - e_{R_1} - e_C; \; i = \frac{\lambda}{L}, \; e_{R_1} = R_1 i \tag{3.16}$$

$$\dot{q} = i - i_2; \; i_2 = \frac{e_{R2}}{R_2}, \; e_C = \frac{1}{C}q, \; e_{R_2} = e_C \tag{3.17}$$

Rearranging Equations (3.16) and (3.17), we obtain

$$\dot{\lambda} = E_{in} - \frac{R_1}{L}\lambda - \frac{1}{C}q \tag{3.18}$$

$$\dot{q} = \frac{1}{L}\lambda - \frac{1}{R_2 C}q \tag{3.19}$$

Q2: Derive the state equation for the electric circuit having a current source shown in Figure 3.9.

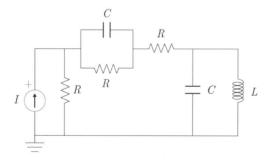

Figure 3.9: An electric circuit composed of several C, R and L components.

3.5   Causality problem in an electronic circuit

Figure 3.10 shows an electric circuit implemented to supply a positive sinusoidal current to a laser diode for modulating the intensity of the laser beam. A laser diode emits a laser beam by the photoelectric principle [9] when the electric current is applied. The ideal voltage source, V_1, is used to supply a sinusoidal current, i_1, to the laser diode, which can be modeled using the resistor as shown in Figure 3.10. The ideal voltage source, V_2, is used to supply a positive DC current, i_3, using a voltage divider comprising R_1 and R_2. i_3 is then added to i_1, producing a positive sinusoidal current, i_5, flowing through the laser diode. Does this circuit work well for adding i_3 to i_1 to produce i_5?

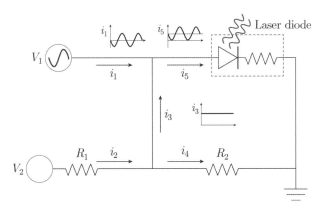

Figure 3.10: An electric circuit implemented to supply a positive sinusoidal current to a laser diode for modulating laser beam intensity.

The corresponding bond graph of the electronic circuit shown in Figure 3.10 is provided in Figure 3.11.

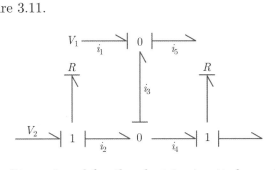

Figure 3.11: Causality assigned for the electric circuit shown in Figure 3.10.

First, we assign the causality marks on the R elements while considering effort sources V_1 and V_2. Effort sources V_1 and V_2 should be denoted using a bar mark in the power direction based on the source causality rule. Next, the bond leaving the 0-junction at the top should have a flow causality denoted by using a point mark in the direction toward the laser diode with the relation, $i_1 + i_3 = i_5$ because it is the design intention of the electric circuit. However, we discover that there is a violation of the causality mark on the 0-junction. There are two bars denoted around the 0-junction at the top. Therefore, this electric circuit does not work as desired.

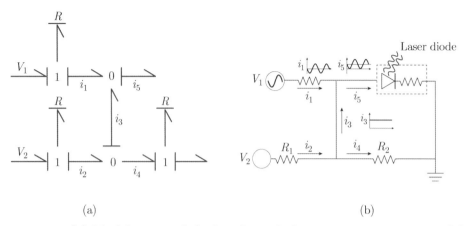

(a)   (b)

Figure 3.12: (a) Modification of the bond graph for correct causality and (b) a real configuration of the electric circuit.

To avoid this causality mark violation, we need to modify the bond graph as shown in Figure 3.12(a). An appropriate configuration of the electric circuit is shown in Figure 3.12(b). A resistor or a capacitor should be inserted after the voltage source as a load to generate current. The causality mark in the R element at the top indicates that the voltage difference is applied to generate current i_1. Without the load, it is not possible to produce i_1. It is meaningful to understand that a load plays the role of producing a desired voltage (effort) or current (flow).

3.6   Bond graph for hydraulic systems

It was stated in Chapter 1 that there are only three parameters in hydraulic systems: fluid inertia I, fluid resistance R, and fluid capacitor C. Moreover, there are four variables, i.e., pressure P, flow rate Q, momentum Γ, and volume V, which are equivalent to force f, velocity v, momentum p, and displacement x in the mechanical system, respectively. The power variables and energy variables of a hydraulic system are listed in Table 1.1.

A bond graph for a passive 1-port of the capacitor, resistor, and inertia can also be introduced in hydraulic systems. Their graphical notations are denoted using the elements C, R, and L. For example, in the C port, when the flow rate is applied through the capacitor, fluid pressure P is then generated using the corresponding constitutive relation in return. In the R port, either flow rate or pressure can be applied, then, pressure or flow rate is respectively generated in return. In the L port, when fluid pressure is applied, the flow rate is then generated using the corresponding constitutive relation in return.

3.6.1 Passive 1-port in hydraulic systems

Fluid resistance, R

Though a fluid system is a type of continuum media, a fluid resistor can be represented as a damper and resistor in mechanical and electric systems, respectively. Figures 3.13(a), (b), and (c) show typical resistors in their most common forms [3]. Figure 3.13(d) shows their corresponding bond graph in which bonds are attached to a 1-junction, indicating how the flow rate is determined in the resistor. They have a constitutive relation relating P_3 to Q_3 as

$$P_3 = f(Q_3) \tag{3.20}$$

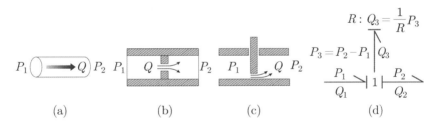

(a)   (b)   (c)   (d)

Figure 3.13: Various fluid resistors (a) short pipe (b) orifice, (c) valve, and (d) their corresponding bond graph.

Figure 3.13(a) shows a short pipe where a viscous force is developed on the inner surface. P_1 and P_2 are the inlet and outlet pressures of the short pipe. When the flow is incompressible and laminar, pressure drop P_3 along the long pipe is expressed based on experimental results [10] as

$$P_3 = \frac{128\mu l}{\pi d^4} Q_3 = R Q_3 \tag{3.21}$$

Hence, the resistance of the long pipe R is represented as

$$R = \frac{128\mu l}{\pi d^4} \tag{3.22}$$

where μ is the viscosity $(N \times s/m^2)$, l is the length, and d is the pipe diameter.

When the flow is turbulent, an approximate expression of Equation (3.20) is [11]

$$P_3 = a_t Q_3 |Q_3|^{3/4} \tag{3.23}$$

where a_t is a constant often determined experimentally. The $P-Q$ relationship is nonlinear; however, it should be noted that long pipes have damping properties because the effort variable is linked to the flow variable. Equation (3.23) indicates that flow rate Q_3 can be determined when there is pressure drop P_3 across the long pipe. Moreover, we can also think when there is a flow rate Q_3 along the long pipe, it causes the pressure drop P_3 because either one is right in the R element.

Figures 3.13(b) and (c) show an orifice and valve where pressure drop P_3 occurs when flow rate Q_3 is produced across the short length. One constitutive relation associated with the R element is [11]

$$P_3 = \frac{\rho}{2(C_d)^2(A_0)^2} Q_3 |Q_3| \tag{3.24}$$

where C_d is the discharge coefficient. A_o is the area of the orifice or the valve. Due to the nonlinear $P-Q$ relationship, R of the orifice and valve are expressed nonlinearly.

Fluid capacitance, C

Figures 3.14(a) and (b) show typical capacitors in their most common forms. Figure 3.14(c) shows their corresponding bond graph in which bonds are attached to a 0-junction, indicating how the pressure is determined in the capacitor. They have constitutive relations relating P_3 to V_3 by

$$P_3 = f(V_3) = f(\int Q_3 dt) \tag{3.25}$$

Figure 3.14(a) shows a tank where the net flow rate made by the in-flow Q_1 subtracted by the out-flow Q_2 is flowing into the tank. P_1 and P_2 are the inlet and outlet pressures of the tank. As the fluid in the tank has potential energy, it can be said that it is modeled as a fluid capacitance C which is obtained from the relation between pressure P_3 and the height filled with fluid h as

$$P_3 = \rho g h = \frac{\rho g}{A} V_3 = \frac{1}{C} V_3 \qquad (3.26)$$

Then, C is expressed from Equation (3.26) as

$$C = \frac{A}{\rho g} \qquad (3.27)$$

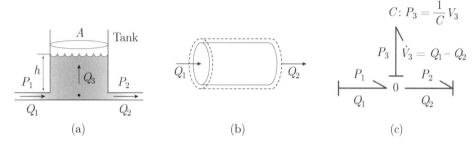

(a)   (b)   (c)

Figure 3.14: Various fluid capacitors (a) tank, (b) flexible pipe, and (c) their corresponding bond graph.

Figure 3.14(b) shows a flexible pipe that can expand to the volume shown by the dotted line due to the integral of net flow rate $\int (Q_1 - Q_3) dt$. It determines the pressure. When the fluid is compressible, it has the same effect as the flexible pipe.

Fluid inertia, I

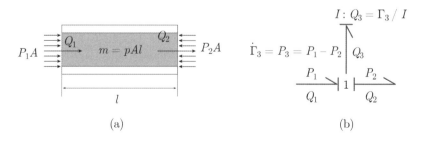

(a)   (b)

Figure 3.15: (a) Fluid inertia in a large pipe and (b) its bond graph.

When the mass effect in the pipe needs to be considered because of a large area, fluid inertia I is modeled with the following constitutive relation:

$$Q_3 = f(\Gamma_3) \tag{3.28}$$

Figure 3.15(a) shows a pipe segment filled with fluid applied by the pressure P_1 and P_2 at both ends. Figure 3.15(b) shows its corresponding bond graph with 1-junction because of the same flow rate through the pipe, i.e., $Q_1 = Q_2$. The pipe segment length is l. To obtain the fluid inertia, we use the force equation because the pressure and flow rate can be converted to force and velocity. For this purpose, we obtain the net force on the fluid, $P_1A - P_2A$, the fluid mass in the volume, ρAl, and the fluid velocity, Q_3/A. Thus, the force equation is obtained by considering the following fluid acceleration:

$$P_1 A - P_2 A = \rho Al \frac{\dot{Q}_3}{A} \tag{3.29}$$

Simplifying Equation (3.29) provides

$$P_3 = \frac{\rho l}{A} \dot{Q}_3 \tag{3.30}$$

Integrating Equation (3.30) and using $\int P_3 dt = \Gamma_3$, we obtain

$$\Gamma_3 = \frac{\rho l}{A} Q_3 = IQ_3 \tag{3.31}$$

Equation (3.31) implies that I is obtained as

$$I = \frac{\rho l}{A} \tag{3.32}$$

3.6.2   State equations of hydraulic systems

As an example, Figure 3.16(a) shows a hydraulic system composed of tanks and short pipes with input Q and output Q_2. Tank 1 and Tank 2 have volumes of V_1 and V_2, respectively. The outlet pressure of Tank 1 and the inlet pressure of Tank 2 are P_1 and P_2, respectively. Tank 1 and tank 2 can be modeled using fluid capacitances Cs based on the description introduced in Section 3.6.1. The short pipes can also be modeled using fluid resistances Rs. Figure 3.16(b) shows its corresponding bond graph.

The state equations can be obtained from the bond graph shown in Figure 3.16(b) as

$$\dot{V}_1 = Q - Q_1; \; P_1 = \frac{1}{C_1}V_1, \; Q_1 = \frac{P_1 - P_2}{R_1}$$

$$\dot{V}_2 = Q_1 - Q_2; \; P_2 = \frac{1}{C_2}V_2, \; Q_2 = \frac{P_2}{R_2}$$

which are rearranged as

$$\dot{V}_1 = Q - \frac{1}{R_1 C_1}V_1 - \frac{1}{R_1 C_2}V_2 \qquad (3.33)$$

$$\dot{V}_2 = \frac{1}{R_1 C_1}V_1 - (\frac{1}{R_1 C_2} + \frac{1}{R_2 C_2})V_2 \qquad (3.34)$$

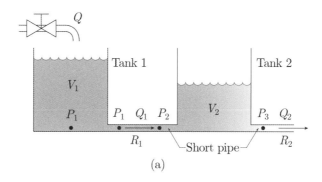

(a)

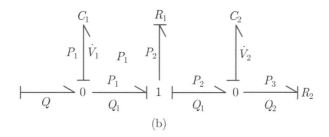

(b)

Figure 3.16: (a) A hydraulic system composed of tanks and valves and (b) its bond graph.

Suppose that the pipe having a resistance R_1 at the end of tank 1 is connected to three pipes having the same resistances, R_2, as shown in Figure 3.17(a). When P_a is the atmospheric pressure, i.e., $P_a = 0$, how much the flow rate is obtained at Pipe 1? This analysis can also be performed using an electric circuit for easy understanding of the hydraulic system, as shown in Figure 3.17(b) in which three R_2 are connected in parallel. Tank 1 is expressed using capacitor C_1. The corresponding bond graph and its simplified bond graph are respectively shown in Figure 3.18(a) and Figure 3.18(b), which considers that the three R_2 connected in parallel can be represented by the equivalent resistance $R_{eq} = \frac{R_2}{3}$.

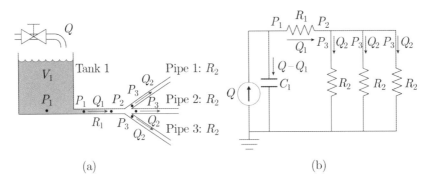

(a)   (b)

Figure 3.17: (a) Schematic and (b) equivalent circuit diagram of a hydraulic system composed of a tank and three pipes.

(a)

(b)

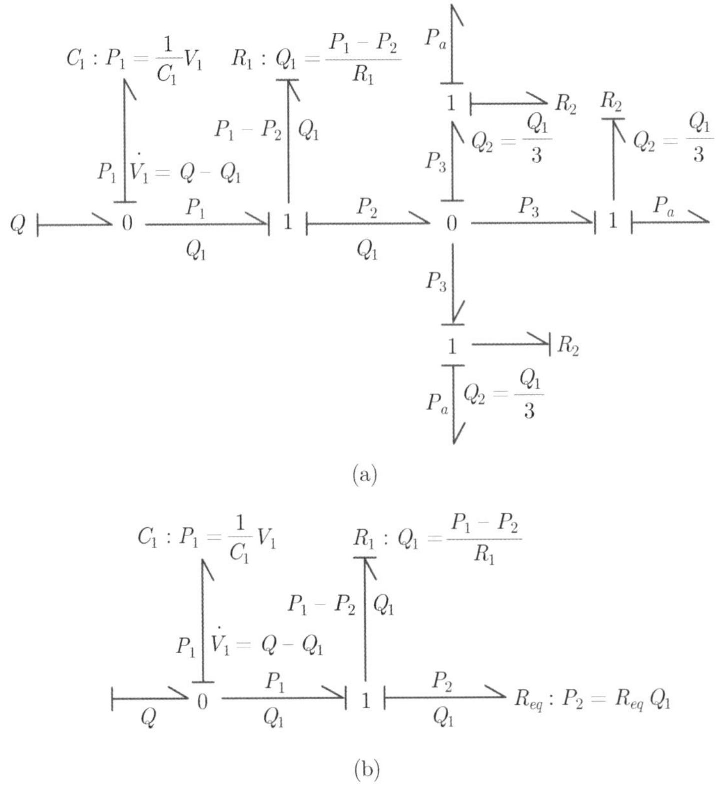

Figure 3.18: (a) Bond graph model of a hydraulic system composed of a tank and three pipes and (b) its simplified bond graph.

The constitutive relations associated with C_1 and R_1 are

$$P_1 = \frac{1}{C_1}V_1 = \frac{1}{C_1}\int(Q-Q_1)dt \qquad (3.35)$$

$$Q_1 = \frac{P_1-P_2}{R_1} \qquad (3.36)$$

We can determine how the pressure and flow rate change in each bond using the constitutive equations when three pipes are used and compare it to the case when one pipe is used. From the electric circuit, three R_2 connected in parallel reduce the load resistance to $\frac{R_2}{3}$. Hence, flow rate Q_1 increases by $3Q_1$ if R_1 is negligibly small. This result affects the constitutive relations associated with R_1 and C_1. The increase of flow rate Q_1 requires P_1 to decrease as expected from Equation (3.35). This conclusion agrees with the law of power conservation, which states that the operating condition changes due to the load effect, as addressed in Section 1.5.2. The first 0-junction represents the practical power source in which pressure decreases when flow increases. Additionally, the increase of flow rate Q_1 requires (P_1-P_2) to increase, as expected from Equation (3.36). Finally, the results state that an increase in Q_1 results in a decrease in P_1 and a greater decrease in P_2, i.e., $P_1 > P_2$. From the above analysis, it is necessary to check that the applied power P_1Q is large enough if we want to supply a desired flow to the pipes.

3.7 Pneumatic systems

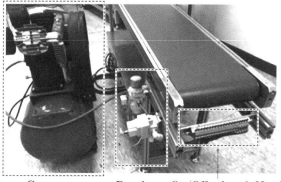

Figure 3.19: Photograph illustrating the configuration of a pneumatic system.

Let us think about a pneumatic system that blows various metal scraps, such as Al, Cu, Pb, and steel, flowing on a conveyor to containers located at different distances from nozzles. This type of system is used for classifying different metal materials by applying the appropriate pneumatic flow rates to metal scraps, as shown in Figure 3.19 [12].

The pneumatic system is constructed using a compressor, regulator, on/off valve, and nozzle, and its power flow is shown in Figure 3.20. When the on-off valve is closed, the maximum pressure is built up at the compressor with zero flow rate. When the on-off valve is open, the flow is determined by the built-up pressure and loads at the nozzle which has a small area. The flow blows a specific metal scrap such that it flies into a container.

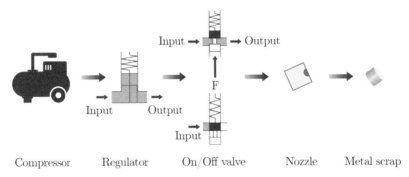

Figure 3.20: Components of the pneumatic system.

An electronic circuit model of the metal classifying the pneumatic system is shown in Figure 3.21(a) and its bond graph is shown in Figure 3.21(b). Here, P_c and Q_c are the supplied pressure and flow rate by the compressor, respectively. The on-off valve is modeled using a switch and the nozzle is modeled using resistor R_n. When the drain pressure is P_1, the flow rate at the nozzle, Q can be determined from the load R_n and the pressure drop $P - P_1$. Suppose that the scrap travel distance is achieved if Q is at the desired flow rate, Q_d. When Q is larger than Q_d, the metal scrap will travel farther than the desired distance. This problem can be solved using a slightly larger R_n so that the metal scrap travels to the desired distance. However, it is practically impossible to find a nozzle having an appropriate R_n on the market. Instead, a regulator can be used to adjust the pressure and flow rate to cope with R_n in the pneumatic system.

Figure 3.22(a) shows a schematic diagram of a regulator modeled using resistor R_{reg} and capacitor C_{reg}, which are represented using a valve and tank as

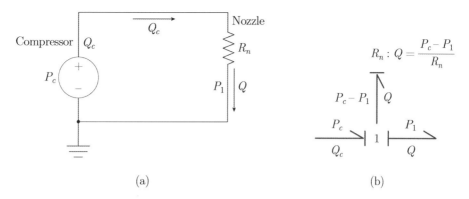

(a)   (b)

Figure 3.21: (a) An electronic circuit model of a pneumatic system with a single nozzle and (b) its bond graph representation.

shown in Figure 3.13(c) and 3.14(a), respectively. Here, P_1 is the pressure of the bottom of the regulator. P_{reg} is the pressure drop between P_c and P_1. Q_{reg} is the difference in flow from the regulator inlet to the regulator outlet.

The corresponding bond graph is represented in Figure 3.22(b). As indicated by the bond graph, R_{reg} and C_{reg} help to adjust the applied pressure and flow rate, P_c and Q_c to the operating pressure and the flow rate, P_1 and Q, respectively.

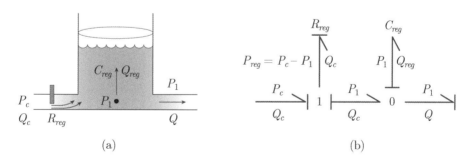

(a)   (b)

Figure 3.22: (a) Schematic diagram of a regulator and (b) its bond graph.

The electronic circuit model of the metal-classifying pneumatic system with a single nozzle having resistance R_n and a regulator is shown in Figure 3.23. Its bond graph is represented in Figure 3.24. The state equation can be obtained from the bond graph. There is just one integral causality for the C element whose energy variable is defined as the volume of the regulator V_{reg}. Therefore,

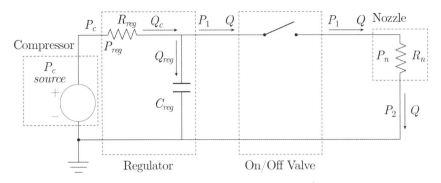

Figure 3.23: Electronic circuit model of a metal-classifying pneumatic system with a single nozzle and regulator.

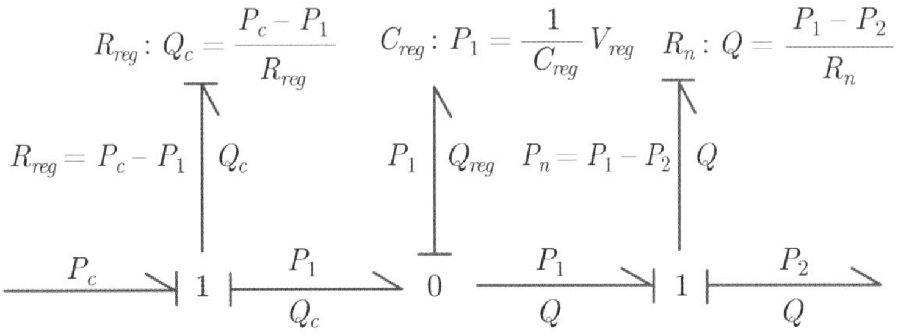

$$R_{reg}: Q_c = \frac{P_c - P_1}{R_{reg}} \quad C_{reg}: P_1 = \frac{1}{C_{reg}} V_{reg} \quad R_n: Q = \frac{P_1 - P_2}{R_n}$$

$$R_{reg} = P_c - P_1 \bigg| Q_c \qquad P_1 \bigg| Q_{reg} \quad P_n = P_1 - P_2 \bigg| Q$$

$$\xrightarrow{P_c} \; 1 \; \Big|\!- \frac{P_1}{Q_c} \; \searrow \; 0 \; \xrightarrow{\;\;} \frac{P_1}{Q} \; \xrightarrow{\;\;} 1 \; \Big|\!- \frac{P_2}{Q} \searrow$$

Figure 3.24: The corresponding bond graph of Figure 3.23.

we have one state equation:

$$\dot{V}_{reg} = Q_c - Q; \quad P_1 \;\; = \;\; \frac{1}{C_{reg}} V_{reg} \tag{3.37}$$

$$Q_c \;\; = \;\; \frac{P_c - P_1}{R_{reg}} \tag{3.38}$$

$$Q \;\; = \;\; \frac{P_1 - P_2}{R_n} \tag{3.39}$$

Rearranging the above relations, we obtain the state equation by assuming $P_2 = 0$ as

$$\dot{V}_{reg} = \frac{P_c}{R_{reg}} - \frac{V_{reg}}{C_{reg}} \left(\frac{1}{R_{reg}} + \frac{1}{R_n} \right) \tag{3.40}$$

Solving Equation (3.40) with initial conditions, we obtain V, P_1, and Q sequentially for given P_c and parameters C_{reg}, R_{reg}, and R_n.

Furthermore, the state equation and the constitutive relations expressed in Equations (3.37), (3.38), and (3.39) can be used to design a metal-classifying pneumatic system for the desired operation. We can simply determine how C_{reg}

and R_{reg} need to be changed without solving Equation (3.40). For example, suppose that the flow rate at the nozzle Q needs to be increased to have a longer travel distance. We notice that Q_c should also be increased from the 0-junction relation at the bond graph represented in Figure 3.24. According to Equation (3.38), we can increase Q_c by decreasing R_{reg}. Alternatively, as an increase in Q_c means a decrease in P_1, according to the law of conservation of power, it is possible to decrease P_1 by increasing C_{reg}, as expected from Equation (3.37).

Multiple nozzles can be used for simultaneously blowing several scraps. The regulator is also used to adjust P_c and Q_c to cope with n nozzles. A pneumatic system with a regulator and multiple nozzles is modeled using an electronic circuit as shown in Figure 3.25. When the same resistance R_n is used for individual nozzles, the equivalent resistance of the multiple nozzles decreases because they are considered resistors connected in parallel. Hence, a higher flow rate, nQ is required too much, the pneumatic system will not work due to the low pressure supplied by the compressor.

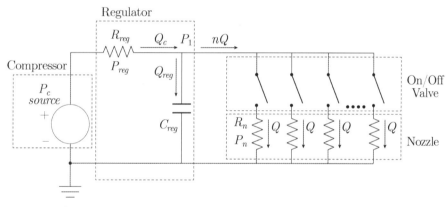

Figure 3.25: An electronic circuit model of a pneumatic system with n nozzles.

The change in the operating condition due to multiple nozzles can be understood using a load curve. When a pneumatic system has just a single nozzle, as shown in Figure 3.23, the load curve is represented using $P_1 = R_n Q$ with the assumption of $P_2 = 0$. Hence, the operating pressure and flow rate can be determined using the constant power curve ($PQ = \text{constant}$) and the load curve, as shown in Figure 3.26. Suppose that the operating condition is determined by $P_1 = (P_1)_s$ and $Q = Q_s$ for a single nozzle due to the regulator. When multiple nozzles are used, the load curve is represented using a lower slope because of the decreased load R_n. Hence, flow rate Q_s is increased to $Q_m = nQ_s$ and pressure

$(P_1)_s$ is decreased to $(P_1)_m$.

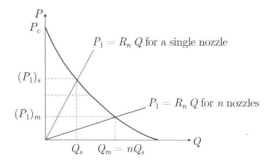

Figure 3.26: Operating pressure and flow rate depending on the pneumatic system load conditions.

Response analysis

Response analysis

4.1   Time response analysis for a first order system

There are several methods for solving differential equations that describes the behavior of a physical system in the time domain [13]. The electrical equation for the circuit shown in Figure 3.2 is represented again in terms of charge as follows:

$$e_{in} = R\frac{dq}{dt} + \frac{1}{C}q \qquad (4.1)$$

Equation (4.1) can also be represented in terms of current for a tangible variable as

$$\frac{de_{in}}{dt} = R\frac{di}{dt} + \frac{1}{C}i \qquad (4.2)$$

Equations (4.1) and (4.2) are first-order differential equations that are solved at two separate conditions: the transient state, which lasts only for a short time, and steady state, which lasts for a long time. Their solutions are commonly called transient and steady state solutions respectively. They are also called homogeneous and particular solutions, respectively. The transient response is obtained by assuming $\frac{de_{in}}{dt} = 0$. The steady state response is obtained by assuming that $\frac{de_{in}}{dt}$ exists. Then, initial conditions are applied to a generalized solution that combines the transient and the steady state responses to obtain a final solution.

At the transient condition, $RC\frac{di}{dt} = -i$ from Equation (4.2). By applying the variable separation method [13], we obtain

$$\frac{di}{i} = -\frac{1}{RC}dt \qquad (4.3)$$

Then, by integrating both sides, we obtain

$$\ln i = -\frac{1}{RC}t + k \tag{4.4}$$

Here, k is an integral constant. Finally, we obtain the transient response as follows:

$$i(t) = e^{(k - \frac{1}{RC}t)} \tag{4.5}$$

At the steady state condition, a more complicated process is usually required to determine i. However, when e_{in} is constant in this case, the steady state response is simply obtained by assuming that $\frac{di}{dt} = 0$ based on the definition of the steady state condition, i.e., i is eventually constant at $t = \infty$. Then, we have $i = 0$ because $\frac{de_{in}}{dt} = 0$. When initial conditions $i(0) = c$ are applied, the final solution of Equation (4.2) is

$$i = ce^{-\frac{1}{RC}t} \tag{4.6}$$

The behavior of the current i with respect to time is shown in Figure 4.1.

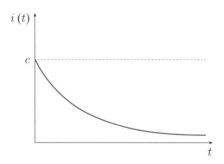

Figure 4.1: Current in an RC circuit when the input voltage is constant.

If $e_{in} = e_o \sin \omega t$, then

$$\frac{de_{in}}{dt} = \omega e_o \cos \omega t \tag{4.7}$$

Using a complicated mathematical process [13], we obtain $i(t)$ as

$$i(t) = ce^{\frac{-t}{RC}} + \frac{\omega e_o C}{1 + (\omega RC)^2}(\cos(\omega t) + \omega RC \sin(\omega t)) \tag{4.8}$$

Figure 4.2 shows the response of $i(t)$ when a sinusoidal input is applied.

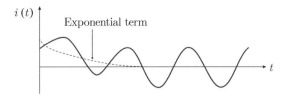

Figure 4.2: Current in an RC circuit when the input voltage is sinusoidal.

We studied that the differential equations of physical systems can be described using first-order state equations by the bond graph modeling in Chapters 2 and 3. Let us now examine how we can obtain the generalized solution of state equations represented in the matrix form of Equation (2.25).

First, for simplicity of analysis, we consider the first-order state equation written in scalar form,

$$\dot{x}(t) - ax(t) = bu \qquad (4.9)$$

If both sides are multiplied by an integrating factor e^{-at}, we obtain

$$e^{-at}\dot{x} - e^{-at}ax = e^{-at}bu \qquad (4.10)$$

The left side of Equation (4.10) is the same as the differential form of

$$\frac{d}{dt}((e^{-at}x(t)) = e^{-at}bu \qquad (4.11)$$

Integrating both sides of Equation (4.11) from 0 to t with respect to a dummy variable τ, we obtain

$$\int_0^t \frac{d}{dt}((e^{-a\tau}x(\tau))d\tau = e^{-at}x(t) - x(0) = \int_0^t e^{-a\tau}bu(\tau)d\tau \qquad (4.12)$$

By rearranging Equation (4.12), we finally obtain

$$x(t) = e^{at}x(0) + \int_0^t e^{a(t-\tau)}bu(\tau)d\tau \qquad (4.13)$$

Further, by expanding the above procedure to a high-order system represented by the matrix form $\dot{\mathbf{X}} = \mathbf{AX} + \mathbf{Bu}$, we can similarly obtain the state response as follows [4]:

$$\mathbf{x}(t) = e^{\mathbf{A}t}\mathbf{x}(0) + \int_0^t e^{\mathbf{A}(t-\tau)}\mathbf{Bu}(\tau)d\tau \qquad (4.14)$$

The state response described in Equation (4.14) consists of two components. The first term is the transient response, which depends on initial conditions $\mathbf{x}(0)$. The second term is the steady state response, which depends on input $\mathbf{u}$. Note that the matrix exponential of matrix $\mathbf{A}$ is defined as [4]

$$e^{\mathbf{A}t} = \mathbf{I} + \mathbf{A}t + \frac{\mathbf{A}^2 t^2}{2!} + \frac{\mathbf{A}^3 t^3}{3!} + \ldots + \frac{\mathbf{A}^k t^k}{k!} + \ldots \tag{4.15}$$

Equation (4.15) is often written using state transition matrix $\Phi(t)$, i.e., $\Phi(t) = e^{\mathbf{A}t}$.

Example 1: Find out the homogeneous response of the state equations with the initial conditions $x_1(0) = 2, x_2(0) = 3$, represented as

$$\dot{x}_1 = -2x_1 + u \tag{4.16}$$
$$\dot{x}_2 = x_1 - x_2 \tag{4.17}$$

Solution 1: System matrices $\mathbf{A}$ and $\mathbf{B}$ are

$$\mathbf{A} = \begin{bmatrix} -2 & 0 \\ 1 & -1 \end{bmatrix}, \quad \mathbf{B} = \begin{bmatrix} 1 \\ 0 \end{bmatrix}$$

By referring to Equation (4.15), the state transition matrix $\Phi(t)$ is

$$\begin{aligned}
\Phi(t) &= e^{\mathbf{A}t} \\
&= (I + \mathbf{A}t + \frac{\mathbf{A}^2 t^2}{2!} + \frac{\mathbf{A}^3 t^3}{3!} + \ldots + \frac{\mathbf{A}^k t^k}{k!} + \ldots) \\
&= \begin{bmatrix} 1 & 0 \\ 0 & 1 \end{bmatrix} + \begin{bmatrix} -2 & 0 \\ 1 & -1 \end{bmatrix} t + \begin{bmatrix} 4 & 0 \\ -3 & 1 \end{bmatrix} \frac{t^2}{2!} + \begin{bmatrix} -8 & 0 \\ 7 & -1 \end{bmatrix} \frac{t^3}{3!} + \ldots \\
&= \begin{bmatrix} 1 - 2t + \frac{4t^2}{2!} - \frac{8t^2}{3!} + \ldots & 0 \\ 0 + t - \frac{3t^2}{2!} + \frac{7t^3}{3!} + \ldots & 1 - t + \frac{t^2}{2!} - \frac{t^3}{3!} + \ldots \end{bmatrix} = \begin{bmatrix} \phi_{11} & \phi_{12} \\ \phi_{21} & \phi_{22} \end{bmatrix}
\end{aligned}$$

Element ϕ_{11} and ϕ_{22} are simply the series representation of e^{-2t} and e^{-t} respectively. Element ϕ_{21} is not so easily recognized; however, it is represented by using $e^{-t} - e^{-2t}$. Hence, $\Phi(t)$ is

$$\Phi(t) = \begin{bmatrix} e^{-2t} & 0 \\ e^{-t} - e^{-2t} & e^{-t} \end{bmatrix} \tag{4.18}$$

Then, the homogeneous responses are

$$
\begin{aligned}
x_1(t) &= x_1(0)e^{-2t} = 2e^{-2t} \\
x_2(t) &= x_1(0)(e^{-t} - e^{-2t}) + x_2(0)e^{-t} = 5e^{-t} - 2e^{-2t}
\end{aligned}
$$

Example 2: Find out the particular response of the state equations described as in Example 1 for an input $u(t) = 5, \quad t > 0$.

Solution 2: The particular response is

$$
\mathbf{x}(t) = \int_0^t e^{\mathbf{A}(t-\tau)} \mathbf{B} u(\tau) d\tau = e^{\mathbf{A}t} \int_0^t e^{\mathbf{A}\tau} \mathbf{B} u(\tau) d\tau \tag{4.19}
$$

Using Equation (4.18), we obtain the following particular response:

$$
\begin{aligned}
\begin{bmatrix} x_1(t) \\ x_2(t) \end{bmatrix} &= \begin{bmatrix} e^{-2t} & 0 \\ e^{-t} - e^{-2t} & e^{-t} \end{bmatrix} \int_0^t \begin{bmatrix} e^{2\tau} & 0 \\ e^{\tau} - e^{2\tau} & e^{\tau} \end{bmatrix} \begin{bmatrix} 5 \\ 0 \end{bmatrix} d\tau \\
&= \begin{bmatrix} e^{-2t} & 0 \\ e^{-t} - e^{-2t} & e^{-t} \end{bmatrix} \begin{bmatrix} \int_0^t 5e^{2\tau} d\tau \\ \int_0^t (5e^{\tau} - 5e^{2\tau}) d\tau \end{bmatrix} \\
&= \begin{bmatrix} \frac{5}{2} - \frac{5}{2}e^{-2t} \\ \frac{5}{2} - 5e^{-t} + \frac{5}{2}e^{-2t} \end{bmatrix}
\end{aligned}
$$

Time domain analysis of dynamic systems represented using differential equations is usually very complicated, as expressed in the above process because the differential equation needs to be solved with specified input. The analysis is further complicated when the system is a higher-order dynamic system. It is almost impossible to obtain an analytical solution in this case. These difficulties lead to frequency analysis of dynamic systems. Now, we find out how frequency analysis is different and how it is beneficial to the system engineer.

4.2 Time response analysis using the Laplace transformation method

The Laplace transformation method [13] can be considered another method for solving differential equations that takes advantage of the operational method.

Applying the Laplace transform to both sides of Equation (4.1), we obtain

$$\mathcal{L}(e_{in}(t)) = \mathcal{L}\left\{R\frac{dq(t)}{dt} + \frac{1}{C}q(t)\right\} = \mathcal{L}\left\{R\frac{dq(t)}{dt}\right\} + \mathcal{L}\left\{\frac{1}{C}q(t)\right\}$$

$$E_{in}(s) = R\mathcal{L}\left\{\frac{dq(t)}{dt}\right\} + \frac{1}{C}Q(s) \tag{4.20}$$

where $E_{in}(s)$ and $Q(s)$ are the Laplace transformed functions of $e_{in}(t)$ and $q(t)$, respectively, which are defined as

$$E_{in}(s) = \mathcal{L}(e_{in}(t)) = \int_0^\infty e^{-st}e_{in}(t)dt$$

$$Q(s) = \mathcal{L}(q(t)) = \int_0^\infty e^{-st}q(t)dt$$

where s is a complex variable. The first term of the right side of Equation (4.20) is obtained using its definition as

$$\mathcal{L}(\frac{dq(t)}{dt}) = \int_0^\infty e^{-st}\frac{dq(t)}{dt}dt = [e^{-st}q(t)]_0^\infty + s\int_0^\infty e^{-st}q(t)dt \tag{4.21}$$

$$= sQ(s) - q(0) \tag{4.22}$$

With the initial condition, $q(0) = 0$, output $Q(s)$ can be obtained using Equations (4.20) and (4.22) as

$$Q(s) = \frac{C}{RCs + 1}E_{in}(s) \tag{4.23}$$

Then, we can obtain the transfer function of $\frac{Q(s)}{E_{in}(s)}$, which is defined as the ratio of the Laplace transformation of the output (response function) to the Laplace transformation of the input (driving function) under the assumption that all initial conditions are zero as follows:

$$\frac{Q(s)}{E_{in}(s)} = \frac{C}{RCs + 1} \tag{4.24}$$

As expressed in Equation (4.24), $\frac{Q(s)}{E_{in}(s)}$ is explicitly expressed and algebraically obtained as a relation of an output versus an input. This is the greatest advantage of the Laplace transformation, through which the dynamic equation in the time domain can be transformed into an algebraic equation in the s complex plane. Then, the relation of an output versus an input can be easily obtained using the transfer function. However, it is impossible to describe the relation between an output versus an input explicitly $\frac{q(t)}{e_{in}(t)}$ in the time domain.

When input $e_{in}(t)$ is given, output $q(t)$ can be obtained using inverse Laplace transforms from $Q(s)$. For example, when $e_{in}(t) = 1$, $q(t)$ is calculated using the relation of $\mathcal{L}(1) = \frac{1}{s}$ and table of the Laplace transforms listed in Table 4.1 as

$$
\begin{aligned}
q(t) &= \mathcal{L}^{-1}\left\{\frac{C}{RCs+1} \cdot E_{in}(s)\right\} = \mathcal{L}^{-1}\left\{\frac{C}{RCs+1} \cdot \frac{1}{s}\right\} \qquad (4.25) \\
&= \mathcal{L}^{-1}\left\{C(\frac{-RC}{RCs+1} + \frac{1}{s})\right\} \qquad (4.26) \\
&= C(1 - e^{-\frac{1}{RC}t}) \qquad (4.27)
\end{aligned}
$$

Table 4.1 shows the most commonly used Laplace transforms. Many other Laplace transforms can be found in [4] and [13].

Table 4.1: Table of commonly used Laplace transforms.

	$f(t)$	$F(s)$
1	Unit impulse $\delta(t)$	1
2	Unit step $1(t)$	$\frac{1}{s}$
3	t	$\frac{1}{s^2}$
4	t^n $(n = 1, 2, 3, 4, \ , \ )$	$\frac{1}{s^n}$
5	e^{-at}	$\frac{1}{s+a}$
6	te^{-at}	$\frac{1}{(s+a)^2}$
7	$\frac{1}{a^2}(1 - e^{-at} - ate^{-at})$	$\frac{1}{s(s+a)^2}$
8	$\frac{1}{a^2}(at - 1 + e^{-at})$	$\frac{1}{s^2(s+a)}$
9	$\sin \omega t$	$\frac{\omega}{s^2+\omega^2}$
10	$\cos \omega t$	$\frac{s}{s^2+\omega^2}$
11	$\frac{1}{a}(1 - e^{-at})$	$\frac{1}{s(s+a)}$
12	$\frac{1}{b-a}(1 - e^{-at} - e^{-bt})$	$\frac{1}{(s+a)(s+b)}$
13	$e^{-at}\sin \omega t$	$\frac{\omega}{(s+a)^2+\omega^2}$
14	$e^{-at}\cos \omega t$	$\frac{s+a}{(s+a)^2+\omega^2}$
15	$\frac{\omega_n}{\sqrt{1-\zeta^2}}e^{-\zeta\omega_n t}\sin(\omega_n\sqrt{1-\zeta^2}t)$	$\frac{\omega_n^2}{s^2+2\zeta\omega_n s+\omega_n^2}$

Let's think again about the electronic circuit shown in Figure 3.2. When the output of the circuit is the voltage across the capacitor $e_c(t)$, $E_c(s)$ is represented as

$$E_C(s) = \frac{1}{C}Q(s) \tag{4.28}$$

Then, the transfer function $\frac{E_c(s)}{E_{in}(s)}$ is easily obtained using Equation (4.23) and Equation (4.28) as

$$\frac{E_C(s)}{E_{in}(s)} = \frac{(1/C)Q(s)}{(Rs + (1/C))Q(s)} = \frac{1}{RCs + 1} \tag{4.29}$$

For example, when $e_{in}(t) = 1$, $E_C(s)$ is obtained as

$$E_C(s) = \frac{1}{s} \cdot \frac{1}{(1 + RCs)} = \frac{1}{s} - \frac{1}{(\frac{1}{RC} + s)} \tag{4.30}$$

Then, $e_c(t)$ is exactly obtained using the inverse Laplace transform as

$$e_C(t) = 1 - e^{-\frac{1}{RC}t} \tag{4.31}$$

The state equations obtained from the bond graph modeling technique introduced in Chapters 2 and 3 can also be Laplace transformed to provide the relation between the output and input. For example, the state equations represented by Equations (3.18) and (3.19) are expressed using the Laplace transform as

$$s\lambda(s) = E_{in}(s) - \frac{R_1}{L}\lambda(s) - \frac{1}{C}Q(s) \tag{4.32}$$

$$sq(s) = \frac{1}{L}\lambda(s) - \frac{1}{R_2C}Q(s) \tag{4.33}$$

where $\lambda(s)$ is the Laplace transform of $\lambda(t)$. Then, the transfer function of the output voltage across the capacitor $E_c(s)$ to the input voltage $E_{in}(s)$ is obtained using Equations (4.32) and (4.33) as

$$\frac{E_c(s)}{E_{in}(s)} = \frac{Q(s)/C}{E_{in}(s)} = \frac{1}{CLs^2 + ((L + R_1R_2C)/R_2)s + (R_1/R_2 + 1)}$$

Q1: Derive the transfer function $\frac{Q_2}{Q}$ of the hydraulic system shown in Figure 3.16 from the state equations represented by Equations (3.33) and (3.34).

4.3 Time response analysis of a second-order system

For a second-order differential equation, system analysis can also be made using the Laplace transform. The system behavior can be understood when a specific input is given. We study one of the most popular mechanical systems composed of a mass, damper, and spring with a force input, as shown in Figure 4.3.

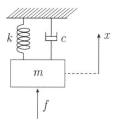

Figure 4.3: A second-order mechanical system with a force input.

The governing equation is represented in second-order differential form as follows:

$$f = m\ddot{x} + c\dot{x} + kx \tag{4.34}$$

where m, c, and k are the mass, damping coefficient, and spring coefficient of the mechanical system. Equation (4.34) is transformed, using the Laplace transformation method under the assumption of $\dot{x} = \ddot{x} = 0$, yielding to

$$F(s) = (ms^2 + cs + k)X(s) \tag{4.35}$$

The transfer function $G(s)$ is defined as the Laplace transform of output displacement $X(s)$ versus the Laplace transform of input force $F(s)$. Then, $G(s)$ is expressed as

$$G(s) = \frac{X(s)}{F(s)} = \frac{1}{ms^2 + cs + k} \tag{4.36}$$

which is rewritten for normalization as

$$\frac{kX(s)}{F(s)} = \frac{k}{ms^2 + cs + k} = \frac{k/m}{\left(s^2 + \frac{c}{m}s + \frac{k}{m}\right)} \tag{4.37}$$

Using more general notations of output $C(s)$ and input $R(s)$, Equation (4.37) can be finally expressed as

$$G(s) = \frac{C(s)}{R(s)} = \frac{\omega_n^2}{s^2 + 2\zeta\omega_n s + \omega_n^2} \tag{4.38}$$

Here, $2\zeta\omega_n = \frac{c}{m}$ and $\omega_n^2 = \frac{k}{m}$. where ω_n is called the natural frequency. ζ is called the damping ratio. Then $\zeta = \frac{c}{2\sqrt{mk}}$.

The dynamic behavior of Equation (4.38) is described based on two parameters: ζ and ω_n. The output $c(t)$ can be obtained by the inverse transform of Equation (4.38) when input $r(t)$ is specified. However, $x(t)$ has different forms depending on the range of ζ, such as under-damped, critically damped, and over-damped responses [4].

Under-damped case ($0 < \zeta < 1$)

In this case, $\frac{C(s)}{R(s)}$ can be written as

$$\frac{C(s)}{R(s)} = \frac{\omega_n^2}{s^2 + 2\zeta\omega_n s + \omega_n^2} = \frac{\omega_n^2}{(s + \zeta\omega_n + j\omega_d)(s + \zeta\omega_n - j\omega_d)} \tag{4.39}$$

where ω_d is called the damped natural frequency, defined as $\omega_d = \omega_n\sqrt{1 - \zeta^2}$. For a unit step input $r(t) = 1$, $R(s) = 1/s$. Then, the partial fraction expansion of $C(s)$ yields

$$
\begin{aligned}
C(s) &= \frac{\omega_n^2}{(s^2 + 2\zeta\omega_n s + \omega_n^2)s} = \frac{1}{s} - \frac{s + 2\zeta\omega_n}{(s^2 + 2\zeta\omega_n s + \omega_n^2)} \\
&= \frac{1}{s} - \frac{s + \zeta\omega_n}{(s + \zeta\omega_n)^2 + \omega_d^2} - \frac{\zeta\omega_n}{(s + \zeta\omega_n)^2 + \omega_d^2}
\end{aligned} \tag{4.40}
$$

The second and third terms of Equation (4.40) are expressed by referring to the inverse Laplace transforms in Table 4.1 as follows:

$$\mathcal{L}^{-1}\left\{\frac{s + \zeta\omega_n}{(s + \zeta\omega_n)^2 + \omega_d^2}\right\} = e^{-\zeta\omega_n t}\cos\omega_d t$$

$$\mathcal{L}^{-1}\left\{\frac{\omega_d}{(s + \zeta\omega_n)^2 + \omega_d^2}\right\} = e^{-\zeta\omega_n t}\sin\omega_d t$$

Hence,

$$c(t) = \mathcal{L}^{-1}\left\{C(s)\right\} = 1 - e^{-\zeta\omega_n t}\left(\cos\omega_d t + \frac{\zeta}{\sqrt{1-\zeta^2}}\sin\omega_d t\right)$$

$$= 1 - \frac{e^{-\zeta\omega_n t}}{\sqrt{1-\zeta^2}}\sin\left(\omega_d t + \tan^{-1}\frac{\sqrt{1-\zeta^2}}{\zeta}\right) \quad (4.41)$$

As written in Equation (4.41), $c(t)$ has steady state and transient responses. It can be seen that the frequency of transient oscillation is the damped natural frequency, w_d, which varies with ζ. It is also noted that $\zeta\omega_n$ should be positive and large to reduce the transient response time and to eventually disappear after a short time.

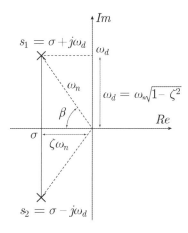

Figure 4.4: Graphical relations of ω_n and ζ.

One thing to note here is that the basic characteristics of a second-order system or plant expressed in Equation (4.38) are included in the characteristic equation, which is defined as

$$\omega_n^2 + 2\zeta\omega_n s + \omega_n^2 = 0 \quad (4.42)$$

The roots of the characteristic equation, s_1 and s_2, are

$$s_{1,2} = -\zeta\omega_n \pm j\omega_d = \sigma \pm j\omega_d \quad (4.43)$$

From the above definitions of σ and ω_d, ω_n and ζ are graphically obtained from

Figure 4.4 as

$$\omega_n = \sqrt{\sigma^2 + \omega_d^2}, \tag{4.44}$$

$$\zeta = cos\beta \tag{4.45}$$

Figure 4.5 shows $c(t)$ expressed by Equation (4.41). It has a transient response exhibiting a damped oscillation before reaching the steady state. The transient response characteristics of the second-order system to a unit step input can be specified by the following features, which are graphically defined in Figure 4.5:

1. Rising time, t_s
2. Peak time, t_p
3. Maximum overshoot, M_p
4. Settling time, t_s

The rising time, t_r, is obtained using the condition $c(t_r) = 1$ and referring to Equation (4.41). The peak time, t_p is the time required for the response to reach the first peak time, which can be obtained using the condition $\frac{dc(t_p)}{dt} = 0$. The maximum overshoot, M_p, is obtained from the condition $M_p = c(t_p)$. The settling time, t_s, is the time required for the response curve to reach and stay within a range about the final value of the size that is specified by an absolute percentage of the final value (usually 2% or 5%) [4].

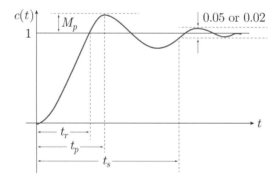

Figure 4.5: Unit step response curve showing several specifications.

The error signal, $e(t)$, is obtained by the difference between input $r(t)$ and output $c(t)$ as follows:

$$
\begin{aligned}
e(t) &= r(t) - c(t) \\
&= \frac{e^{-\zeta \omega_n t}}{\sqrt{1 - \zeta^2}} \sin\left(\omega_d t + \tan^{-1} \frac{\sqrt{1 - \zeta^2}}{\zeta}\right)
\end{aligned}
$$

When ζ equals zero, $c(t)$ is obtained from Equation (4.41) as

$$
c(t) = 1 - \cos \omega_n t
$$

The output response for the zero-damping case becomes undamped and oscillation continues infinitely.

Critically damped case ($\zeta = 1$)

If $\zeta = 1$, the two poles of $C(s)/R(s)$ are equal, yielding $s = -\omega_n$ from Equation(4.43), and the system may be approximated as a critically damped. For a unit-step input, $R(s) = 1/s$ and $C(s)$ can be written as

$$
C(s) = \frac{\omega_n^2}{(s + \omega_n)^2 s} \tag{4.46}
$$

The inverse Laplace transform of Equation (4.46) is found to be

$$
c(t) = 1 - e^{-\omega_n t}(1 + \omega_n t) \quad (t \geq 0) \tag{4.47}
$$

This result can be directly obtained using the following mathematical relationship from Equation (4.41), under the condition that ζ converges to 1, as follows:

$$
\lim_{\zeta \to 1} \frac{\sin \omega_d t}{\sqrt{1 - \zeta^2}} = \lim_{\zeta \to 1} \frac{\sin(\omega_n \sqrt{1 - \zeta^2} t)}{\sqrt{1 - \zeta^2}} = \omega_n t
$$

Over-damped case ($\zeta > 1$)

In this case, the two poles of $C(s)/R(s)$ are negative real and unequal. For a unit-step input, $R(s) = 1/s$ and $C(s)$ can be written as

$$
C(s) = \frac{\omega_n^2}{(s + \zeta\omega_n + \omega_n\sqrt{\zeta^2 - 1})(s + \zeta\omega_n - \omega_n\sqrt{\zeta^2 - 1})s} \tag{4.48}
$$

The inverse Laplace transform of Equation (4.48) is

$$c(t) = 1 + \frac{1}{2\sqrt{\zeta^2 - 1}(\zeta + \sqrt{\zeta^2 - 1})}e^{-(\zeta + \sqrt{\zeta^2 - 1})\omega_n t}$$
$$- \frac{1}{2\sqrt{\zeta^2 - 1}(\zeta + \sqrt{\zeta^2 - 1})}e^{-(\zeta - \sqrt{\zeta^2 - 1})\omega_n t}$$
$$= 1 + \frac{\omega_n}{2\sqrt{\zeta^2 - 1}}\left(\frac{e^{-s_1 t}}{s_1} - \frac{e^{-s_2 t}}{s_2}\right) \quad (t \geq 0) \tag{4.49}$$

where $s_1 = (\zeta + \sqrt{\zeta^2 - 1})\omega_n$ and $s_2 = (\zeta - \sqrt{\zeta^2 - 1})\omega_n$. Thus, response $c(t)$ includes two decaying exponential terms.

The output responses, $c(t)$, are shown for under-damped, critically damped, and over-damped cases with different damping ratios, ζ in Figure 4.6.

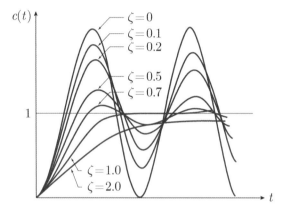

Figure 4.6: The output responses of a second-order system with different damping ratios, ζ.

As addressed above, the time domain analysis provides an accurate result. However, it is a very complicated and tedious process because the differential equation needs to be solved. Moreover, output responses should be calculated again when an input changes. Therefore, it can be said that time domain analysis is not a good tool for understanding the system's behavior in an engineering sense.

4.4 Frequency response analysis

4.4.1 Experimental frequency response

Instead of time domain analysis, frequency response analysis is used because it quickly provides a large amount of information regarding system behavior characteristics for various inputs without a need to solve the differential equations. The frequency response is the steady state response of a system to a sinusoidal input with respect to frequency. This information allows one to obtain the output response in terms of magnitude and phase differences compared with the input. It can be directly obtained from the transfer function represented by the Laplace transformation, which is an additional advantage for system analysis. To easily understand this method, an experimental frequency response is first investigated before an analytical frequency response which will be introduced in the next section.

Consider a linear system whose transfer function is $G(s)$, as shown in Figure 4.7. Then, $G(s)$ is represented as follows:

$$G(s) = \frac{Y(s)}{X(s)} \tag{4.50}$$

Figure 4.7: A linear system whose transfer function is $G(s)$.

Let us suppose that the input and output of a plant or system are denoted by $x(t)$ and $y(t)$, respectively. If a sinusoidal input signal $x(t)$ having frequency ω is applied to the plant, the output will also be a sinusoidal signal with the same frequency but with a possibly different magnitude and phase angle. Then, the mathematical description of $x(t)$ is

$$x(t) = X \sin \omega t \tag{4.51}$$

When ω varies from low to high frequency, $y(t)$ can be obtained with the different magnitudes and phase delays and can be described as

$$y(t) = Y \sin(\omega t + \phi) \tag{4.52}$$

For example, Figure 4.8 shows input and output signals that are experimentally obtained at a specific frequency. The output sinusoidal signal has a magnitude reduction ratio $M = 1/2$ and phase delay $\phi = -80°$ compared to the input signal.

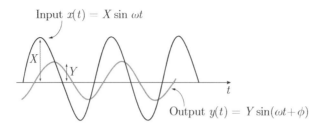

Figure 4.8: Input and output sinusoidal signals with different amplitude and phase delay.

The magnitude ratio, $M = \frac{Y}{X}$, and the phase delay ϕ are experimentally obtained for all frequencies and plotted using solid dots, as shown Figure 4.9. They are called the frequency response. M and ϕ are important indexes for analyzing the dynamic characteristics or specifications of an electrical circuit or mechanical plant because they provide information regarding how well the output follows the input.

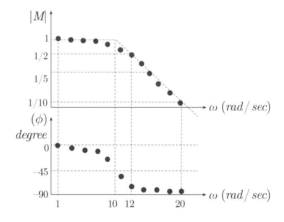

Figure 4.9: Experimentally obtained frequency response.

Figure 4.9 indicates that the output follows well the input at a low frequency because $M \cong 1$ and $\phi \cong 0$. As the frequency increases, the output does not follow well the input in terms of magnitude reduction and phase delay. When the frequency is higher than 10 rad/sec, there is a severe magnitude reduction and phase delay in the output. Since the frequency response shows how the output behaves for the sinusoidal input, we can figure out how well the plant is capable of following the input. For example, when the robot arm moves an object to a target position using the torque generated from an electric motor in the robot arm, input X and output Y are considered the torque and the rotating angle, respectively. According to its frequency response, the input force frequency should be much lower than 10 rad/sec to exhibit a better tracking performance of the robot arm to the target position.

However, if Figure 4.9 is the frequency response of the angle output Y of the robot arm for an external vibration input X from the ground, motion slower than 10 rad/sec causes vibration in the robot arm motion because the vibration input is transmitted without reduction, i.e., $Y = X$. Therefore, the frequency response can be applied to the analysis and design of a dynamical system because it provides much information related to the system behavior characteristics without resorting to complicated differential equations.

Additionally, the experimental frequency response is also useful when it is difficult to determine an analytical plant model due to the unexpected dynamics in high frequency or modeling uncertainties. These problems are severe, especially in mechanical systems. In the above cases, the experimental frequency response can be applied to accurately identify the system model.

As an example, an experiment is performed to obtain the frequency response of the electrical system shown in Figure 3.2. The input is the supplied voltage, e_{in}, and the output is the voltage, e_C, across capacitor C. The electrical system can be implemented on a breadboard using a resistor R and a capacitor C, which are properly selected to provide a desired transfer function. A function generator can be used to apply the sinusoidal input voltage with varying magnitude and frequency. An oscilloscope is used for measuring the input and output signals to determine the magnitude reduction ratio and phase delay. Figure 4.10 shows a configuration of the experimental frequency response test. Here, V_{++} and *Ground* are the points to connect to the + voltage supply and the ground in the power supply.

It is convenient to apply a sinusoidal input signal with a constant magnitude for all frequencies. For larger numerical representation, we can plot the magnitude ratio with respect to the frequency on a logarithmic scale instead of a linear scale. This frequency response is called a Bode plot.

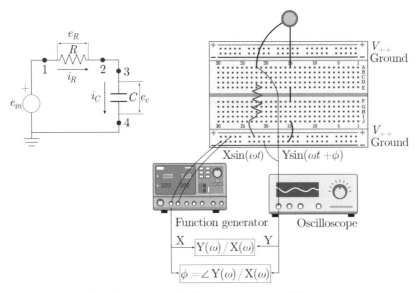

Figure 4.10: Configuration of the experimental frequency response test.

4.4.2 Analytical frequency response

It was proven that the frequency response of a plant can also be obtained directly from its transfer function, $G(s)$, with s being replaced by $j\omega$ [4]. Thus, we obtain the following relations:

$$|G(j\omega)| = \left|\frac{Y(j\omega)}{X(j\omega)}\right| = M \tag{4.53}$$

$$\angle G(j\omega) = \angle \frac{Y(j\omega)}{X(j\omega)} = \phi \tag{4.54}$$

Hence, when $G(s)$ can be derived or obtained from the system modeling process, the dynamic behavior of various systems can be analytically obtainable.

Proof: Suppose that $G(s)$ can be written as a ratio of two polynomials in s, i.e.,

$$G(s) = \frac{p(s)}{q(s)} = \frac{p(s)}{(s + s_1)(s + s_2) \cdots (s + s_n)}$$

For a stable system, $s_1, s_2, \cdots s_n$ are positive numbers. The output $Y(s)$ is then

$$Y(s) = G(s)X(s) = \frac{p(s)}{q(s)}X(s) \tag{4.55}$$

where $X(s)$ is the Laplace transform of $x(t) = X \sin \omega t$ and is $\frac{\omega X}{s^2 + \omega^2}$. Thus, $Y(s)$ is

$$
\begin{aligned}
Y(s) &= G(s)X(s) = G(s)\frac{\omega X}{(s^2 + \omega^2)} \\
&= \frac{a}{s + j\omega} + \frac{\bar{a}}{s - j\omega} + \frac{b_1}{s + s_1} + \frac{b_2}{s + s_2} + \cdots + \frac{b_n}{s + s_n} \quad (4.56)
\end{aligned}
$$

where a and b_i (where $i = 1, 2, ..., n$) are constant and $\bar{a}$ is the complex conjugate of a. Then, $y(t)$ is obtained by the inverse transform of Equation (4.56) as

$$
y(t) = ae^{-j\omega t} + \bar{a}e^{j\omega t} + b_1 e^{-s_1 t} + b_2 e^{-s_2 t} + \cdots + b_n e^{-s_n t} \quad (4.57)
$$

As t approaches infinity, all terms on the right side drop out except for the first two terms. Therefore, the steady-state response, $y_{ss}(t)$ is

$$
y_{ss}(t) = ae^{-j\omega t} + \bar{a}e^{j\omega t} \quad (4.58)
$$

a and b can be evaluated using Equation (4.56) as follows:

$$
\begin{aligned}
a &= G(s)\frac{\omega X}{s^2 + \omega^2}(s + j\omega)|_{s=-j\omega} = -\frac{XG(-j\omega)}{2j} \\
\bar{a} &= G(s)\frac{\omega X}{s^2 + \omega^2}(s - j\omega)|_{s=j\omega} = \frac{XG(j\omega)}{2j}
\end{aligned}
$$

Since $G(j\omega)$ is a complex quantity, it can be written as

$$
G(j\omega) = |G(j\omega)|e^{j\phi} \quad (4.59)
$$

Similarly, $G(-j\omega)$ can be written as

$$
G(-j\omega) = |G(j\omega)|e^{-j\phi} \quad (4.60)
$$

Then, Equation (4.58) can be rewritten using Equations (4.59) and (4.60), yielding

$$
\begin{aligned}
y_{ss}(t) &= -\frac{X|G(j\omega)|e^{-j\phi}}{2j}e^{-j\omega t} + \frac{X|G(j\omega)|e^{j\phi}}{2j}e^{j\omega t} \\
&= X|G(j\omega)|\frac{e^{j(\omega t + \phi)} - e^{-j(\omega t + \phi)}}{2j} \quad (4.61)
\end{aligned}
$$

Finally, Equation (4.61) is arranged by noting $e^{j(\omega t + \phi)} = \cos(\omega t + \phi) + j\sin(\omega t + \phi)$ as

$$
\begin{aligned}
y_{ss}(t) &= X|G(j\omega)|\sin(\omega t + \phi) \\
&= Y\sin(\omega t + \phi)
\end{aligned}
$$

where $Y = X|G(j\omega)|$. From the above statement, the magnitude ratio of the output to the input M and phase delay ϕ are directly obtained from $|G(j\omega)|$ and $\angle G(j\omega)$.

4.4.3 Frequency response of a first-order system

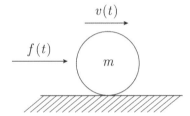

Figure 4.11: Schematic of a mechanical system with mass.

Consider a simple mechanical system composed of mass m, as shown in Figure 4.11. When the input and output are applied force $f(t)$ and the velocity $v(t)$, we obtain the following mechanical equation: $f(t) = m\frac{dv}{dt}$. Hence, $v(t)$ is obtained by integrating $\frac{f(t)}{m}$. A graphical representation of this relation is shown in Figure 4.12(a) using a block diagram when $m = 1$. Suppose $F(s)$ and $V(s)$ are the Laplace transforms of $f(t)$ and $v(t)$, respectively. Since the Laplace transformation of time integration, $\int f(t)dt$ is expressed as $F(s)/s$, the transfer function $G(s)$ of the system shown in Figure 4.11 can be represented in Figure 4.12(b). Hence, $1/s$ is called an integrator.

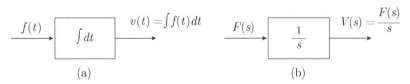

Figure 4.12: (a) Time integration of $f(t)$ and (b) Laplace transformation of time integration of $f(t)$.

Of course, $G(s)$ can be directly obtained from the first order mechanical equation $f(t) = m\frac{dv}{dt}$ as follows:

$$G(s) = \frac{V(s)}{F(s)} = \frac{1}{ms} \qquad (4.62)$$

Then, $G(j\omega)$ is

$$G(j\omega) = \frac{1}{jm\omega} \qquad (4.63)$$

M is obtained using a logarithmic magnitude of $G(j\omega)$ with the base of the logarithm of 10. The unit of this representation is the decibel (db). For the sake of scale representation,

$$M(j\omega) = 20\log|G(j\omega)| = 20\log|\frac{1}{jm\omega}| = -20\log(m\omega)\ (db) \qquad (4.64)$$

$$\phi(j\omega) = \angle(\frac{1}{jm\omega}) = -90° \qquad (4.65)$$

M and ϕ are described in Figure 4.13, assuming $m = 1$. M is plotted on a logarithmic scale Bode plot with respect to applied frequency ω. According to Equation(4.64), M is reduced by 20 db for one decade increase of frequency. There is a constant phase delay of $-90°$, regardless of frequency.

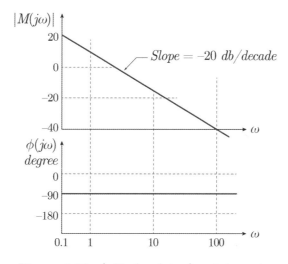

Figure 4.13: A Bode plot of an integrator.

Q2: Draw the frequency response of the transfer function $G(s) = \frac{1}{0.01s}$ in a logarithm scale. Sketch the output for the inputs $x = \sin 2\pi(10Hzt)$

and $x = \sin 2\pi(100Hzt).$

Consider again the electrical circuit shown in Figure 3.2. When the input and output are $E_{in}(s)$ and the capacitor voltage $E_C(s)$, respectively, the transfer function of the electrical circuit $G(s)$ is obtained from the first-order differential equation associated with the RC circuit as

$$\frac{E_C(s)}{E_{in}(s)} = \frac{\frac{1}{Cs}I(s)}{(R + \frac{1}{Cs})I(s)} = \frac{1}{1 + RCs} \tag{4.66}$$

Then, $G(j\omega)$ is

$$G(j\omega) = \frac{1}{1 + RCj\omega} \tag{4.67}$$

Amplitude ratio M and phase delay ϕ of $G(j\omega)$ are respectively obtained as

$$M = 20\log|G(j\omega)| = 20\log\left|\frac{1}{1 + jRC\omega}\right| = -20\log\sqrt{1 + (RC\omega)^2} \tag{4.68}$$

$$\phi = \angle(\frac{1}{1 + jRC\omega}) = -\tan^{-1}RC\omega \tag{4.69}$$

At a low frequency, such that $\omega \ll \frac{1}{RC}$, M is approximated by

$$M = -20\log\sqrt{1 + (RC\omega)^2} \cong -20\log 1 = 0 \ (db) \tag{4.70}$$

Thus, the log-magnitude curve at a low frequency is the constant $0 \ db$ line. The phase delay ϕ is approximated by

$$\phi = \angle(\frac{1}{1 + jRC\omega}) \cong -\tan^{-1}0 = 0° \tag{4.71}$$

At a high frequency, such that $\omega \gg \frac{1}{RC}$, the log magnitude is approximated by

$$M = -20\log\sqrt{1 + (RC\omega)^2} \cong 20\log\sqrt{(RC\omega)^2} = -20\log RC\omega \ (db) \tag{4.72}$$

The phase delay ϕ is approximated by

$$\phi = \angle(\frac{1}{1 + jRC\omega}) \cong -\tan^{-1}\infty = -90° \tag{4.73}$$

Finally, M and ϕ are described in Figure 4.14 in a logarithmic scale Bode plot with respect to applied frequency ω. The frequency at which the $0 \ db$ curve and -20 *db/decade* curve meet is called a cut-off frequency or corner frequency, ω_c.

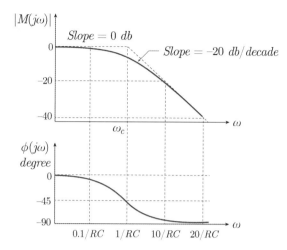

Figure 4.14: Frequency response of a first-order electrical system.

Based on the frequency response, we know the magnitude of the output voltage, $e_C(t)$, is almost the same as that of input voltage, e_{in}, and there is little phase delay at low frequencies. However, the magnitude of output voltage $e_C(t)$ is reduced and there is a phase delay of $-45°$ at the cut-off frequency. The magnitude of $e_C(t)$ is greatly reduced, with a slope of -20 *db/decade*, and there is a phase delay of $-90°$ at high frequencies; hence, it is called a low pass filter (LPF), which is interpreted as a "low-frequency passing" or "following" device. It can be used to reduce the magnitude of high-frequency signals. A typical application of the LPF is electronic noise reduction because noise mainly appears at high frequencies. A high-order LPF can be constructed to further reduce noise magnitude at high frequencies even though a larger phase delay is generated.

Q3: Draw the frequency response of the transfer function $G(s) = \frac{1}{1+0.01s}$ in a Bode plot. Sketch the output for input $x = \sin 2\pi (10\,Hz\,t)$ and $x = \sin 2\pi (100\,Hz\,t)$. (Note that $1\,Hz = 2\pi\ rad/sec$)

Q4: Explain why a constant force is not integrated at a low frequency for a plant whose transfer function is $\frac{1}{s+100}$.

Q5: If we are asked to design an electronic LPF that can obtain the displacement from velocity signal $\sin(100\,Hz\,t)$ which order of LPF is required? How big a cutoff frequency is determined? What will happen if an ideal integrator$(\frac{1}{s})$ is used instead of an LPF when a disturbance, such as DC offset, exists in the electronic circuit?

Now, when the output is the voltage across the resistor, e_R, in the electrical circuit shown in Figure 3.2, the transfer function $G(s)$ is obtained from the first-order differential equation associated with the RC circuit as follows:

$$G(s) = \frac{E_R(s)}{E_{in}(s)} = \frac{R}{R + 1/(Cs)} = \frac{RCs}{1 + RCs} \tag{4.74}$$

Then, $G(j\omega)$ is represented as

$$\frac{E_R(j\omega)}{E_{in}(j\omega)} = \frac{jRC\omega}{1 + jRC\omega} \tag{4.75}$$

We can draw $G(j\omega)$ directly by plotting it with respect to ω ranging from a low to high frequency. Another easy way to draw $G(j\omega)$ is that we separately plot the composed terms, $jRC\omega$ and $\frac{1}{1+jRC\omega}$, and add individual curves graphically in the Bode plot because the logarithm of the multiplied terms is equivalent to adding the logarithms of each term. This is an advantage of the logarithmic plot of the Laplace transformation.

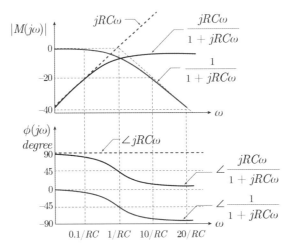

Figure 4.15: Frequency response of a high pass filter.

Differently from the integrator representation $\frac{1}{s}$, s is the Laplace transform of the time differentiation of $\frac{d}{dt}$. Hence, $jRC\omega$ is called a differentiator having a gain of RC. This gain increases the magnitude of the output proportionally with respect to ω with a slope of 20 $db/decade$. As we know that $\frac{1}{1+jRC\omega}$ is the first-order LPF, Equation (4.75) can be easily drawn by adding $jRC\omega$ to $\frac{1}{1+jRC\omega}$, as shown in Fig 4.15.

The transfer function, $G(s)$, represented by Equation (4.75) has features wherein the output magnitude is much reduced at a low frequency, while it is constant with a large magnitude at a high frequency; hence, it is called a high pass filter (HPF), which is interpreted as a "high-frequency passing" or "following" device. It can be used to reduce the magnitude of low-frequency signals, such as the electrical offset or DC (constant) disturbance, which mainly occurs at a low-frequency range.

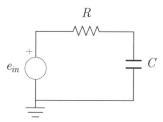

Figure 4.16: A simple electronic circuit composed of resistor R and capacitor C.

When an electronic circuit is designed, it is sometimes required to know its impedance in order to know its load effect. The impedance of a circuit can be easily obtained using the transfer function. For example, for a circuit composed of a resistor R and capacitor C, as shown in Figure 4.16, the equation of the electronic circuit is represented by

$$e_{in} = Ri + \frac{1}{C}\int i\,dt = (R + \frac{1}{Cs})I \tag{4.76}$$

where e_{in} is the input voltage and i is the current. When $E_{in}(s)$ and $I(s)$ are the Laplace transforms of e_{in} and i, respectively, the impedance Z of the circuit is obtained by its definition $\frac{E_{in}(s)}{I(s)}$ as

$$Z = R + \frac{1}{C\omega} \tag{4.77}$$

As represented in Equation (4.77), Z is dependent on frequency. When a high-frequency alternating voltage input is applied, the load due to capacitor C is negligibly small. However, when a low-frequency alternating voltage input is applied, the load due to capacitor C is so large that no current can flow in the circuit.

4.4.4   Frequency response of a second-order system

Now, we consider an electrical circuit composed of a resistor, inductor, and capacitor (RLC), as shown in Figure 4.17, when the input is e_{in}. If the voltage of the capacitor, e_c, is an output, transfer function $G(s)$ of the output to input is obtained from a differential equation of the RLC circuit and is represented by

$$G(s) = \frac{E_C(s)}{E_{in}(s)} = \frac{1/Cs}{Ls + R + 1/(Cs)} = \frac{1/LC}{s^2 + (R/L)s + 1/LC} \qquad (4.78)$$

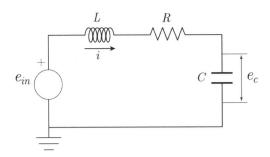

Figure 4.17: A RLC circuit.

The standard forms of Equation (4.78) is written using ζ and ω as follows:

$$G(s) = \frac{E_C(s)}{E_{in}(s)} = \frac{\omega_n^2}{s^2 + 2\zeta\omega_n s + \omega_n^2} \qquad (4.79)$$

where ζ and ω_n are defined respectively by $\frac{R}{2}\sqrt{\frac{C}{L}}$ and $\sqrt{\frac{1}{LC}}$. An analogous system of the RLC electrical circuit is found in the mechanical system shown in Figure 4.3, whose transfer function is expressed using Equation (4.36) and is identical to that of the electrical system expressed using Equation (4.79). This result is expected from the same bond graph modeling. Therefore, the analysis obtained from the transfer function of an electrical system can be also applied to a mechanical system and vice versa.

From Equation (4.79), the frequency response function $G(j\omega)$ is obtained using general notations of input $X(j\omega)$ and output $Y(j\omega)$ as follows:

$$G(j\omega) = \frac{Y(j\omega)}{X(j\omega)} = \frac{\omega_n^2}{\omega_n^2 - \omega^2 + j(2\zeta\omega_n\omega)} \tag{4.80}$$

Then, amplitude ratio M and phase delay ϕ of $G(j\omega)$ are respectively obtained as

$$M = |G(j\omega)| = \left| \frac{1}{1 - (\frac{\omega}{\omega_n})^2 + j2\zeta\frac{\omega}{\omega_n}} \right| = \left| \frac{1}{a + bj} \right| \tag{4.81}$$

$$\phi = \angle(\frac{1}{a + bj}) \tag{4.82}$$

where $a = 1 - (\frac{\omega}{\omega_n})^2$ and $b = 2\zeta\frac{\omega}{\omega_n}$.

The amplitude ratio M and phase delay ϕ of $G(j\omega)$ are respectively rewritten as

$$M = 20\log|G(j\omega)| = 20\log\left|\frac{1}{a + jb}\right| = -20\log\sqrt{a^2 + b^2}\ (db) \tag{4.83}$$

$$\phi = -\tan^{-1}(\frac{b}{a}) \tag{4.84}$$

At a low frequency, such that $\omega \ll \omega_n$, $a \cong 1$ and $b \cong 0$. Hence, M is

$$M = -20\log\sqrt{a^2 + b^2} \cong -20\log 1 = 0\ (db) \tag{4.85}$$

Thus, the log-magnitude curve at low frequencies is the constant 0 db line. Similarly, phase delay ϕ is

$$\phi = \angle(\frac{1}{a + bj}) \cong -(\frac{1}{1 + 0j}) = 0° \tag{4.86}$$

At a high frequency, such that $\omega \gg \omega_n$, $a \cong -(\frac{\omega}{\omega_n})^2 \doteq -\infty$. b is also infinity; however, it is small compared to a because ω is much smaller than ω^2. Thus, b can be approximated to zero, and M is given by

$$M = -20\log\sqrt{a^2 + b^2} \cong -20\log\sqrt{a^2} \cong -20\log(\frac{\omega}{\omega_n})^2 = -40\log\frac{\omega}{\omega_n}\ (db) \tag{4.87}$$

and ϕ is

$$\phi = \angle(\frac{1}{a + bj}) \cong -(\frac{1}{-\infty}) \cong -180° \tag{4.88}$$

Let us now examine how M and ϕ change when $\omega = \omega_n$. For this case, $a = 0$ and $b = 2\zeta$. Hence, M is represented by

$$M = -20 \log \sqrt{a^2 + b^2} = -20 \log(2\zeta) \; (db) \qquad (4.89)$$

and ϕ is represented by

$$\phi = \angle\left(\frac{1}{a + bj}\right) = \angle\left(\frac{1}{0 + j\, 2\zeta}\right) \qquad (4.90)$$

As represented by Equations (4.89) and (4.90), M and ϕ are dependent on ζ. It is recognized that when ζ is very small, M is much amplified and ϕ approaches $-90°$. Based on the above analysis, M and ϕ can be described for a different damping ratio ζ with respect to $\frac{\omega}{\omega_n}$ in Figure 4.18.

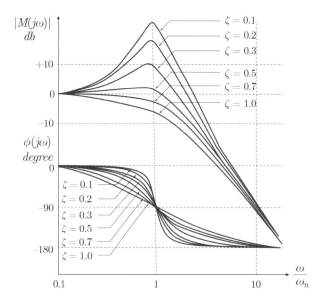

Figure 4.18: Frequency response of a second-order system.

The frequency responses of first-order and second-order systems, shown in Figures 4.14 and 4.18, respectively, exhibit the general characteristics. No significant change is observed in the magnitude ratio at low frequencies, whereas a large reduction is observed at high frequencies; therefore, they are called low-pass filters (LPFs). It is interesting to note again that a mechanical system composed of a mass, spring, and damper is also nothing more than an LPF, such as an electrical system composed of an inductor, capacitor, and resistor. The difference between the first-order and the second-order systems is that there is a peak in the second-order system which depends on ζ, whereas there is no peak in the first-order system. The peak occurs when applied frequency ω is approximately equal to the natural frequency of the system ω_n, which is called the resonant frequency. Additionally, the second-order system has a larger magnitude reduction and phase delay compared to the first-order system.

Q6. Draw the time response of the second-order system represented by Equation (4.79) at $\omega = 100Hz$ when $\omega_n = 100Hz$ and $\zeta = 0.7$.

Q7: How does the displacement of a mechanical system composed of a mass and damper behave when a sinusoidal force is applied to a unit mass?

Q8: Draw the Bode plot for the following transfer function::

$$G(s) = \frac{10,000(s+10)}{(s^2 + 100s + 10,000)}$$

When the output is the displacement of the mass, $x(t)$, it is obtained through double integration of force $f(t)$ because the transfer function $\frac{X(s)}{F(s)}$ is represented by $\frac{1}{s^2}$. If a constant force 1 N is applied, the output displacement continues to increase because of the integral of the input, which eventually makes the mechanical plant collide toward one side. When a sine function is applied to the integrator, a negative sine signal will be produced due to double integration. However, a real output signal has drift motion which causes instability because of a low frequency or DC (constant) disturbance included in the applied force. Hence, caution is required when an integrator is included in a mechanical system.

Example 3: A sinusoidal force $f(t)$ is applied to an unit mass with an additional constant disturbance input of 0.1 N. Show your work how the output displacement x behaves.

Solution 3: The transfer function of the plant, $G(s)$, is represented by $\frac{1}{s^2}$ the frequency response shown in Figure 4.19(a). The sinusoidal inputs $\sin t$ and constant input 0.1 are separately considered for applying to the mass. Then, the outputs are obtained by double integrating the individual inputs. Thus, the mathematical description of $x(t)$ is

$$x(t) = \frac{0.1t^2}{2} - \sin(t) \tag{4.91}$$

The result is shown in Figure 4.19(b).

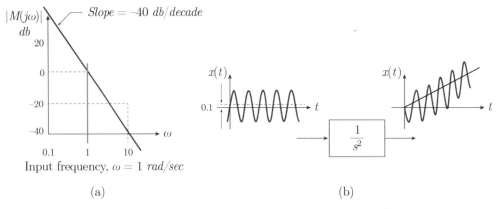

Input frequency, $\omega = 1 \ rad/sec$

(a)   (b)

Figure 4.19: (a) Frequency response of the double integrator $\frac{1}{s^2}$ and (b) output displacement of the double integrator for constant and sinusoidal inputs,

As can be seen from time response $x(t)$, a plant with an integrator has drift motion and will eventually collide toward one side. Therefore, such a plant is difficult to use without feedback control.

4.4.5   Specifications of steady-state response characteristics

We investigated the time response characteristics of the second-order system, such as peak time, maximum overshoot, and settling time in Section 4.3. There are also similar characteristics that can be obtained mathematically in the frequency response.

If $G(j\omega)$ has a peak value at $\omega = w_r$, the frequency is called the damped resonant frequency. A peak value of $|G(j\omega)|$ occurs when the denominator of the second term of Equation (4.81),

$$g(w) = (1 - \frac{w^2}{w_n^2})^2 + (2\zeta\frac{w}{w_n})^2 \tag{4.92}$$

is a minimum. Thus, the resonant frequency w_r is obtained when the frequency at which $\frac{dg}{d\omega}|_{\omega=\omega_r} = 0$ as

$$\omega_r = \omega_n \sqrt{1 - 2\zeta^2} \tag{4.93}$$

Then, the magnitude of the resonant peak M_r can be found by substituting Equation (4.93) into Equation (4.81). For $0 \leq \zeta \leq 0.707$,

$$M_r = |G(jw)|_{max} = |G(jw_r)| = \frac{1}{2\zeta\sqrt{1-\zeta^2}} \tag{4.94}$$

The system bandwidth, ω_{bw}, is generally defined as the frequency at which the magnitude of the transfer function is $+3$ db or -3 db. Then, we have two bandwidths, ω_{bw1} and ω_{bw2} as shown in Figure 4.20. If ω_{bw} is defined to indicate the maximum frequency for which the output follows well the input signal with a small magnitude ratio reduction and small phase delay, ω_{bw1} is more appropriate system bandwidth than ω_{bw2} from a control point of view. However, if the phase delay is not of concern, ω_{bw2} can also be called as the system bandwidth from a signal processing point of view. Based on Figure 4.20, we can say that ω_{bw1} and ω_{bw2} are approximately equal to the resonant frequency, i.e., $\omega_{bw} = \omega_r$. Assuming $\omega_r \cong \omega_n$, we can say $\omega_{bw} \cong \omega_n$. In conclusion, when a system or plant has a high natural frequency, the output signal follows the high-frequency input signal well. Therefore, the bandwidth of the system is considered a speed performance indicator.

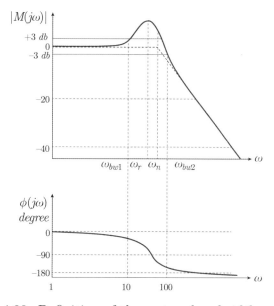

Figure 4.20: Definition of the system bandwidth ω_{bw}.

Q9: We can find real examples of a plant whose behavior can be understood through frequency response analysis. What do you think about the bandwidth of an automobile driven at a distance of 100 m along a curved road, as shown in Figure 4.21 with a speed of 100 km/hr?

Figure 4.21: Illustration of a curved road traveled by an automobile.

Q10: What is the bandwidth of an airplane landing on the ground with a speed of 360 km/hr and landing displacement of 1 km? The landing trajectory is shown in Figure 4.22.

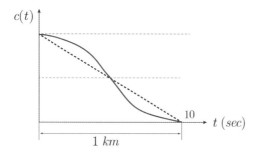

Figure 4.22: The landing trajectory of an airplane landing on the ground.

4.4.6 Engineering applications of frequency response analysis

It is possible to obtain the output response using the frequency response for different inputs in the time domain. Suppose that there are systems or plants which have low and high natural frequencies, as shown in Figures 4.23(a) and (b), respectively. When a sinusoidal command input is applied to the systems, what are the output responses for different systems? For example, when $x(t) = \sin(100Hzt)$ is applied to the system shown in Fig 4.23(a), output $y(t)$ has magnitude reduction of -20 db with a phase delay of $-180°$, as shown in Figure

4.24(a). However, when $x(t)$ is applied to the system shown in Fig 4.23(b), $y(t)$ has little magnitude reduction with a phase delay of $-20°$, as shown in Figure 4.24(b).

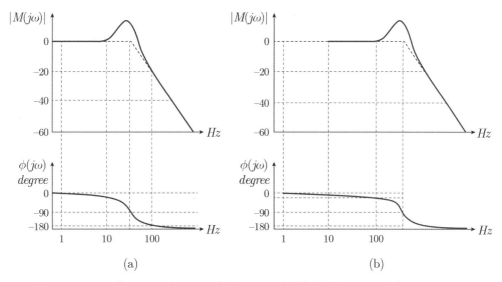

Figure 4.23: Systems having (a) low and (b) high natural frequencies.

From the above examples, it is easily understood that the output cannot follow a high-frequency input (high speed) effectively when the system has a low natural frequency. However, the output can follow a high-frequency input (high speed) effectively when the system has a higher natural frequency. This result shows how a high-speed system should be designed. As can be seen from the output response results for different natural frequencies shown in Figure 4.24(a) and (b), a higher natural frequency ω_n is required for a high-speed system.

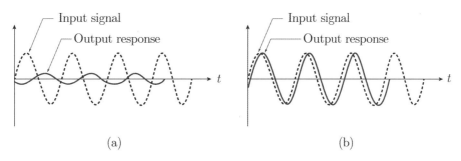

Figure 4.24: Output responses corresponding to the systems shown in (a) Figure 4.23(a) and (b) Figure 4.23(b).

In addition to the command input signal, the system input includes other signals such as disturbance and electronic noise. It is important to recognize their dominant frequency ranges. Fortunately, they belong to different frequency ranges. Particularly, a command input signal mainly exists at a low frequency, ω_{com}, whereas noise mainly exists at a high frequency, ω_{noise}, which are illustrated in Figure 4.25(a) and Figure 4.25(b) in time and frequency domains, respectively.

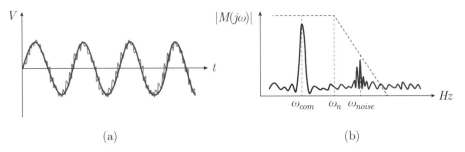

(a)   (b)

Figure 4.25: (a) Signal contaminated with noise in the time domain and (b) signal and noise frequency spectrum.

Q11: Investigate the output position signal when a low-frequency command signal (or DC offset) and high-frequency noise are applied to a plant with a mass and damper system?

If the input is not a command signal but a disturbance or noise, such as mechanical vibration and electrical noise, we need to approach the situation differently. As the disturbance or noise should be prevented from entering a plant, they should be reduced as much as possible. Hence, the system needs to be designed to have a low natural frequency because the disturbance or noise can be reduced at a high frequency. Therefore, it is necessary to properly adjust ω_n of a mechanical or electrical system according to the input types so that only the command input signal survives and the noise is filtered out.

In addition, we can use frequency response analysis for designing mechanical systems. When input x_c is the cam displacement and output x_f is the follower displacement, the transfer function of the cam-follower device shown in Figure

1.11 is represented as

$$\frac{X_f(s)}{X_c(s)} = \frac{k_2}{m_f s^2 + cs + (k_1 + k_2)}$$

$$= \frac{k_2/m_f}{s^2 + 2\zeta\omega_n s + \omega_n^2}$$

$$= \frac{k_2}{(k_1 + k_2)} \cdot \frac{\omega_n^2}{s^2 + 2\zeta\omega_n s + \omega_n^2}$$

where $\omega_n^2 = (k_1 + k_2)/m_f$ and $\zeta = c/2(k_1 + k_2)$. Then, the frequency response is

$$\frac{X_f(j\omega)}{X_c(j\omega)} = \frac{k_2}{(k_1 + k_2)} \cdot \frac{\omega_n^2}{\omega_n^2 - \omega^2 + j2\zeta\omega_n\omega}$$

$$= \frac{k_2}{(k_1 + k_2)} \cdot \frac{1}{1 - (\frac{\omega}{\omega_n})^2 + j2\zeta\frac{\omega}{\omega_n}} \qquad (4.95)$$

This is similar to the frequency response of the second-order system represented in Figure 4.18 except that it has a magnitude ratio of $\frac{k_2}{(k_1+k_2)}$. As the design goal of the cam follower device is $x_f(t) = x_c(t)$, it is evident, according to its frequency response, to design the device so that it has high ω_n, which yields to have a high bandwidth and $\frac{k_2}{(k_1+k_2)} \cong 1$. Therefore, it is concluded that m_f should be low and k_2 should be large compared with k_1.

Q12: An earthquake seismograph shown in Figure 1.14 aims to record vibration signals due to an earthquake.
a) Derive the governing equation of the seismograph when the input is ground vibration x_g due to an earthquake and the output is mass displacement x_m. Re-drive the governing equation of the seismograph when the input is x_g and the output is the relative displacement x_r defined by the difference of x_m and x_g.
b) What is the transfer function of the seismograph when the input is x_g and the output is x_r?
c) Suggest a good design requirement for the earthquake seismograph in terms of x_g and x_r.
d) Describe how a seismograph should be designed to meet the design requirement using the frequency response of the plant.

4.5 Steady state response analysis for various time inputs

We used the frequency response to obtain the steady-state time response when a sinusoidal input is applied. What would happen when the input is not sinusoidal? Can we expect how a plant or system behaves using a frequency response if the inputs, for example, include an abrupt change in magnitude, such as a step or impulse? The answer is yes because such inputs can also be represented by the summation of sinusoidal inputs using Fourier series representation [13].

If $x(t)$ is a period function of period T, $x(t)$ can be represented by a trigonometric series according to the Fourier Series theorem as

$$x(t) = a_0 + \Sigma_{n=1}^{\infty}(a_n \cos(nt) + b_n \sin(nt)) \tag{4.96}$$

$$a_0 = \frac{1}{T} \int_{-T/2}^{T/2} x(t)dt \tag{4.97}$$

$$a_n = \frac{1}{\pi} \int_{-T/2}^{T/2} x(t) \cos(nt)dt \tag{4.98}$$

$$b_n = \frac{1}{\pi} \int_{-T/2}^{T/2} x(t) \sin(nt)dt \tag{4.99}$$

Here, the mathematical expression a_0 is the mean (or average) of $x(t)$. Hence, $x(t)$ is composed of the mean (average) value, a_0, and high-order frequency signals.

For example, a unit step input $x(t)$ can be represented using Equations (4.97), (4.98), and (4.99). As a result, $a_0 = 0$, $a_n = 0$, and $b_n = \frac{\pi}{4}[\sin t, \frac{\sin(3t)}{3}, \frac{\sin(5t)}{5}.......]$. Figure 4.26 graphically represents that unit step input $x(t)$ is approximately the same as the summed signal of sinusoidal functions whose frequencies are ω_1, ω_2, and ω_3. Here, ω_1 is called the fundamental frequency. The summation of much higher frequencies can lead to a correct shape of the unit step input $x(t)$ signal.

Then, what is the output response for a step input when a plant is a second-order mechanical system with natural frequency ω_n and damping ratio 0.5? Suppose that the frequency of applied step input, $\omega_s = \frac{2\pi}{T}$ is lower than the natural frequency of a plant, ω_n. Then, even though ω_s is lower than ω_n, the abrupt change of input magnitude occurred at $T = -\frac{2}{T}$, 0, and $\frac{2}{T}$ produces signals containing ω_n and higher frequencies. Therefore, the step input consists

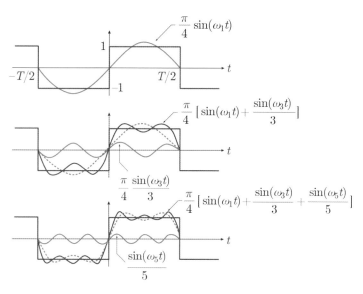

Figure 4.26: A unit step input represented by the summation of sinusoidal inputs having ω_1, ω_2, and ω_3.

of several sinusoidal inputs whose frequencies are ω_1, ω_2, ..., ω_m, ..., ω_n, ..., and ω_∞ as shown in the left part of Figure 4.27. When ω_m is assumed to be much lower than ω_n, the output follows well those input signals whose frequencies are ω_1, ω_2, ..., ω_m based on frequency response analysis. Input signals whose frequencies are close to ω_∞ that are much higher than ω_n do not contribute much to the output response due to the -40 $db/decade$ magnitude reduction rate. However, input signals close to ω_n can produce a large amplitude because of the resonance effect. Therefore, the output response is dominantly affected by input signals whose frequencies are close to ω_n. As a result, the output response at $t > 0$ is obtained by summing all responses that can be individually obtained from the frequency responses at ω_1, ω_2, ..., ω_m, ..., ω_n, ..., and ω_∞, as shown in the right part of Fig 4.27.

The output response is drawn again in Figure 4.28 to relate the important time specification to the frequency specification. M_p, t_p, and t_s indicated in Figure 4.28 are the maximum overshoot, peak time, and settling time, respectively. Though these performance specifications can be correctly determined by the mathematical approach from $y(t)$, as shown in Figure 4.5, it is graphically obtained with conceptual insight with little mathematical error. We can determine a relation between t_p and the damped natural frequency, ω_d from the frequency of the curve of the first part of time response $y(t)$, i.e., $2t_p = \frac{2\pi}{w_d}$.

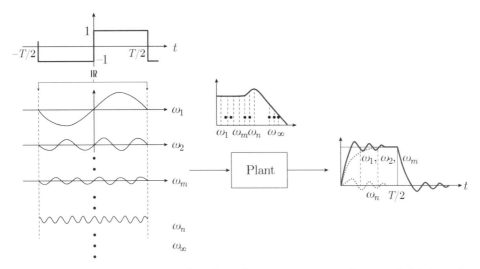

Figure 4.27: Output response of a plant for a step input when ω_n is higher than ω_s.

Hence, we have

$$t_p = \frac{\pi}{w_d} \qquad (4.100)$$

The next step is to investigate an output response for a step input when ω_s is

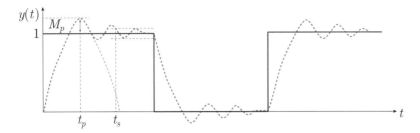

Figure 4.28: Output response of the second-order system for a step input based on frequency response analysis.

higher than ω_n. In this case, the fundamental frequency, ω_1 is higher than ω_n. Hence, the output does not follow well to input signals whose frequencies are ω_1, $\omega_2, \omega_3, \cdots, \omega_\infty$ except ω_n, because it produces a large amplitude and phase delay. Therefore, the output response is dominantly obtained by the response obtained from ω_n, as shown in Fig 4.29.

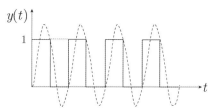

Figure 4.29: Output response $y(t)$ when ω_s is higher than ω_n.

What will the output response be like if $x(t)$ is the unit impulse function, as shown in Figure 4.30? The unit impulse function occurring at $t = 0$ is defined as

$$\delta(t) = 0 \quad \text{for } t \neq t_o$$
$$\delta(t) = \infty \quad \text{for } t = t_o$$
$$\int_{-\infty}^{\infty} \delta(t)dt = 1$$

Since the Laplace transform of the unit impulse function is 1, i.e., $X(s) = 1$,

Figure 4.30: A unit impulse function.

we obtain

$$Y(s) = G(s)$$
$$Y(j\omega) = G(j\omega),$$

which indicates that $y(t)$ is the same as $g(t)$ obtained by the inverse transform of the transfer function $G(s)$. If we plot the input, $X(s) = 1$ in the frequency domain, we know that $X(j\omega)$ has a constant magnitude over a wide frequency from low to high, meaning that the impulse input is theoretically equivalent to applying all frequencies to a plant. Therefore, when an impulse input is applied to a plant experimentally, the plant can be identified from the output response. In other words, the transfer function of the plant is obtained by exciting it with a unit impulse and measuring the output response.

Q13: Figures 4.31(a) and (b) show the frequency responses of two different plants. When the ramp input shown in Figure 4.31(c) is applied to the plants, how do the outputs behave for the different plants in the time domain? Provide valid reasons for your answer.

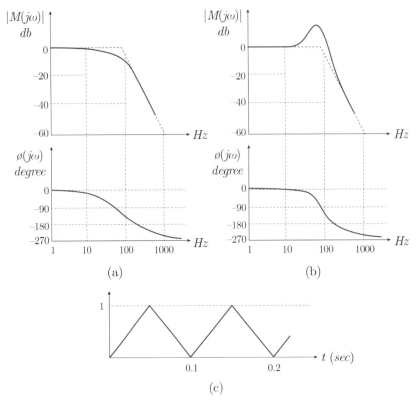

(a)

(b)

(c)

Figure 4.31: Frequency responses of a plant (a) without an overshoot, (b) with an overshoot, and (c) a ramp input applied to the plants.

When a desired input is applied to a plant or system, it is important not to include signals in the input that can cause resonance in the plant. However, though step or ramp inputs are easy to generate, they cause resonance which provides oscillation in systems, as shown in Figure 4.27. This is because these inputs contain a signal whose frequency is close to the resonant frequency of the plant. Therefore, it is necessary to create an input profile w hose f requency is far away from the resonant frequency of physical systems.

To not cause the oscillation in a system or plant due to the resonant frequency, the step input signal shown in Figure 4.32(a) can be modified so that it does not have an abrupt change, as shown in Figure 4.32(b). Here, the input profile consists of a portion of the line with constant magnitude and a portion of the curve in the corner. The former component can be treated as a low-frequency signal. The latter component can be approximated mathematically by a sinusoidal signal that fits into the shape of the corner. If the frequency of this input is less than the resonant frequency of a plant, the resonance effect can be largely reduced because the frequency is the highest among the signals composing the input signal. Other vibration reduction methods can also be found in [14] and [15].

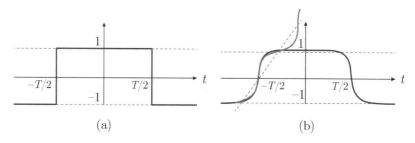

Figure 4.32: (a) Step input signal and (b) input signal without an abrupt change.

4.6 Transient response based on frequency response analysis

It was assumed that frequency response analysis was performed under a steady state condition when an input is applied as discussed in Section 4.4.2. Hence, the initial conditions, such as $\dot{x}(0)$ and $\ddot{x}(0)$ were disregarded at the frequency response because they are associated with a transient response. Mathematically, the above statement is expressed using a second-order system composed of a mass, damper, and spring, for example, with force input F and displacement output x as

$$(ms^2 + cs + k)X(s) = F(s) \tag{4.101}$$

What will happen when there is no input in the second-order system? Particularly, since the input force F is zero, the second-order system expressed by Equation (4.101) is rewritten as

$$(ms^2 + cs + k)X(s) = 0 \tag{4.102}$$

Since $F = 0$, we know that $x(t)$ will be zero at a steady state. However, motion is caused by the initial conditions. This response is called a transient response which differs from the forced response where there is a force to drive the motion. The characteristic of a transient response is that the motion will die out at the steady state, which is why it is called a transient response.

Frequency response analysis can be expanded to investigate how a transient response behaves though it is considered only for a steady state condition. For example, when a mass is pressed at a distance of 1 cm and then released, the initial conditions are, $x(0-) = 0.01\ m$ m and $x(0+) = 0$. These initial conditions can be understood as an abrupt change occurring from $x(0-) = 0.01\ m$ m and $x(0+) = 0$ as the time changes from $t = 0-$ to $t = 0+$. The step input due to the abrupt displacement change can be represented by the summation of sinusoidal signals from Fourier transformation. Among the sinusoidal signals, there must be a resonant frequency of the second-order system. Therefore, the resonant effect is observed in the transient response, as shown in 4.33.

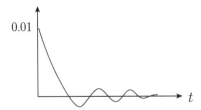

Figure 4.33: Resonant effect observed in a transient response due to abrupt initial conditions.

For example, the frequency response for the transient condition is used for investigating the output response of the system shown in Figure 4.31(b). The applied ramp input can be approximately decomposed into a low-frequency pseudo-sinusoidal input and high-frequency input with a sharp edge due to the abrupt change, as shown in Figure 4.34. When a 10 Hz low-frequency pseudo-sinusoidal input is applied, the output response can be obtained through the frequency response analysis of the steady state, yielding a small magnitude reduction and phase delay. However, the high-frequency input of the sharp edge can be considered a transient input because it is applied discontinuously. This input can include a signal of the resonance frequency, yielding a large oscillation and phase delay in the output response. Therefore, the final output response is obtained by combining the steady state and transient responses, as shown in Figure 4.34.

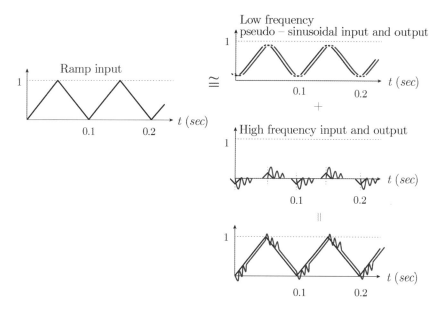

Figure 4.34: An applied ramp input that is approximately decomposed of a low-frequency pseudo-sinusoidal input and a high-frequency input.

4.7   Frequency response of a high-order system

We have studied the characteristics of the first-order and second-order linear systems. However, a system can be higher than the second order. Yet, a third-order system can be decomposed into first- and second-order systems in terms of transfer function representation. A fourth-order system can be decomposed into two second-order systems. Therefore, a higher-order system has characteristics similar to first- and second-order systems.

Figure 4.35(a) shows a typical frequency response of a second-order system that has a fundamental resonance. Hence, the output signal can be excited by the fundamental natural frequency of the system, ω_n when a step input is applied, as shown in Figure 4.35(b).

However, a higher-order system has a frequency response with sub-resonances in addition to the fundamental resonance, as shown in Figure 4.36(a). Hence, the output signal can be excited by sub-resonant frequencies, ω_{s1} and ω_{s2} as well as ω_n, as shown in Figure 4.36(b). The resonant peak due to ω_n can be reduced using feedback control. As a result, the oscillation can be reduced. However, it is almost impossible to reduce the oscillation due to ω_{s1} and ω_{s2} because their frequencies are so high that their resonant peaks are usually difficult to control

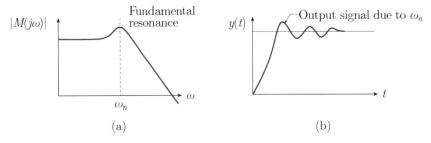

Figure 4.35: (a) Frequency response of a second-order system having a fundamental resonance at ω_n and (b) output signal excited by ω_n in the time domain.

using feedback control. In conclusion, a system having the frequency response shown in Fig 4.36(a) should be redesigned to have no sub-resonances or to have sub-resonances at much higher frequencies to reduce the peaks.

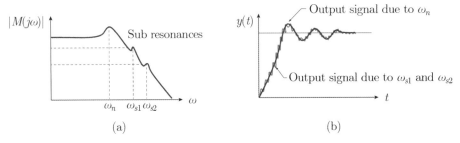

Figure 4.36: (a) Frequency response of a higher-order system having sub-resonances at ω_{s1} and ω_{s2} and (b) output signal excited by ω_{s1} and ω_{s2} in the time domain.

Electromechanical systems

Electromechanical systems

To understand an electromechanical system, it is necessary to understand the origins of modern theories of electricity and magnetism, which are unified into an electromagnetic theory [17]. Hence, the subject of electromechanics necessarily includes the study of electromagnetic fields. It is often necessary to calculate the magnetic flux or density of a magnetic system because they are related to the electromagnetic force. Several analytical or numerical methods have been used to calculate the electromagnetic force [18],[19],[20]. However, it is also meaningful to know the physical concepts associated with the magnetic flux or density of a magnetic system. Therefore, we must study the fundamental equations based on the Maxwell equations. Note that it is necessary to understand field density variables, such as B, H, D, and E, for magnetic and electric system analyses because they are field systems and not lumped systems. In this chapter, we also study how electrical system variables are related to magnetic and mechanical system variables.

5.1 Magnetic field system

The electromagnetic field and source quantities are related by the following partial differential equations [17]:

$$\nabla \times \mathbf{H} = \mathbf{J}_f \tag{5.1}$$

$$\nabla \cdot \mathbf{B} = 0 \tag{5.2}$$

$$\nabla \cdot \mathbf{J}_f = 0 \tag{5.3}$$

$$\mathbf{B} = \mu_o \mathbf{H} \tag{5.4}$$

and

$$\nabla \times \mathbf{E} = -\frac{\partial \mathbf{B}}{\partial t} \tag{5.5}$$

Here, $\mathbf{H}$ and $\mathbf{B}$ are the magnetic field intensity and magnetic flux density, respectively; $\mathbf{J}_f$ is the current density denoted in continuum theory; $\mathbf{E}$ is the electric field intensity; and μ_o is the magnetic permeability of free space, which indicates how well the magnetic flux is generated in the air, and is equal to $4\pi \times 10^{-7} H/m$. Even though there are time-varying sources, $\mathbf{H}$ and $\mathbf{B}$ are determined by considering the system as magneto-static. Then, $\mathbf{E}$ is obtained from the resulting flux density using Equation (5.5). This is the origin of the term "quasi-static magnetic field system" [17]. Please refer to mathematical books or references for definitions of algebra operators, such as ∇, curl operator $\times$, and divergence operator $\cdot$.

Based on Maxwell's equations, we understand that $\mathbf{H}$ is generated in the curl direction when current is applied to a coil. As a result, $\mathbf{B}$ is generated in proportion to $\mathbf{H}$ with a corresponding permeability. If we only refer to the equations, it looks like $\mathbf{H}$ is generated first and $\mathbf{B}$ is consequently produced. However, it is easier to understand magnetic systems when $\mathbf{B}$ is produced first by an applied current and H is the intensity required to supply $\mathbf{B}$ in the air or a material [20].

The partial differential equations described by Equations (5.1) to (5.5) can also be represented in integral form. Physical concepts are easier to understand via the integral forms of Maxwell's equations. Before we derive the integral forms, we first study Stoke's theorem and the divergence theorem [13].

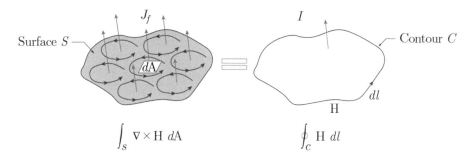

$$\int_S \nabla \times \mathrm{H} \; d\mathrm{A} \qquad\qquad \oint_C \mathrm{H} \; dl$$

Figure 5.1: Graphical illustration of Stoke's theorem.

As shown in Figure 5.1, the integral of the curl of $\mathbf{H}$ ($\nabla \times \mathbf{H}$) over the surface S can be considered the integral of each segment composed of the magnetic flux generated by $\mathbf{J}_f$ flowing through an infinitesimally small area vector $d\mathbf{A}$. Since the magnetic fluxes adjacent to each other inside the contour C are canceled, except for those in the contour, we can understand Stoke's theorem as the integral of the total amount of swirl ($\nabla \times \mathbf{H}$) can be calculated by integrating around the edge, which is represented by

$$\int_S \nabla \times \mathbf{H} \, d\mathbf{A} = \oint_C \mathbf{H} \cdot d\mathbf{l} \qquad\qquad (5.6)$$

Here, S is the surface enclosed by contour C. If we take an integral of Equation (5.1) over surface S, we obtain

$$\int_S \nabla \times \mathbf{H} \, d\mathbf{A} = \int_S \mathbf{J}_f \, d\mathbf{A} = I \qquad\qquad (5.7)$$

Here, I is the current flowing through surface S. From Equations (5.7) and (5.6), the final integral form of Equation (5.1) is

$$\oint_C \mathbf{H} \cdot d\mathbf{l} = \int_S \mathbf{J}_f \, d\mathbf{A} = I \qquad\qquad (5.8)$$

This relation is called Ampere's law [21].

The divergence theorem indicates that the integral of the divergence of $\mathbf{B}$, $\nabla \cdot \mathbf{B}$ over the volume is equal to the value of the function at the surface which bounds the volume. This theorem can be understood by the fact that the source producing $\mathbf{B}$ inside the volume is the same as the amount of $\mathbf{B}$ radiating through the surface. The divergence theorem can be graphically understood using the illustration shown in Figure 5.2.

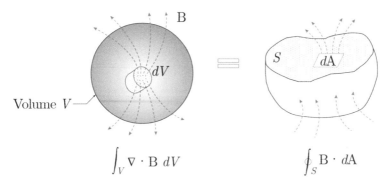

$$\int_V \nabla \cdot \text{B} \, dV \qquad\qquad \oint_S \text{B} \cdot d\text{A}$$

Figure 5.2: Graphical illustration of the divergence theorem.

Based on the divergence theorem, we obtain the following relation:

$$\int_V \nabla \cdot \mathbf{B} \, dV = \oint_S \mathbf{B} \cdot d\mathbf{A} = 0 \tag{5.9}$$

Here, V is the volume enclosed by area S.

Similarly, we have the integral form of Equation (5.5). By applying Stoke's theorem, the surface integral of $\nabla \times \mathbf{E}$ is obtained via line integral $\mathbf{E}$. Thus, we obtain

$$\oint_C \mathbf{E} \cdot d\mathbf{l} = -\frac{d}{dt} \int_S \mathbf{B} \cdot d\mathbf{A} = -\frac{d\phi}{dt} \tag{5.10}$$

Here, ϕ is the magnetic flux flowing through surface S defined as $\int_S \mathbf{B} \cdot d\mathbf{A}$. Equation (5.10) indicates that the electric field represented by $\mathbf{E}$ is related to the magnetic field represented by $\mathbf{B}$, which is referred to as Faraday's law [22]. We will further discuss this relation later. For reference, magnetic flux ϕ has the unit of $Weber$ (Wb). 1 Wb is equal to 10^4 G $(gauss)$. The magnetic flux density, B, has the unit of $Tesla$ (T) defined by Wb/m^2.

5.2   Analysis of the magnetic field system

5.2.1   Magneto-static system

When current I is applied to the N-turn long air-core solenoid shown in Figure 5.3, a magnetic flux is generated. By applying Equation (5.8), we obtain

$$\oint_C \mathbf{H} \cdot d\mathbf{l} = Hl = NI \tag{5.11}$$

where l is the length of the flux line. H is composed of components in the radial, circumferential, and height directions, denoted as H_r, H_ϕ, and H_z, respectively. First of all, does H_r exist? Suppose that H_r is positive when the current flows, as shown in Figure 5.3. If the current direction is reversed, H_r would be negative. However, reversing the current direction is equivalent to turning the solenoid upside down, and it certainly should not change the direction of H_r. Hence, the answer is $H_r = 0$. How about H_ϕ? Applying Ampere's law, which is represented by Equation (5.8), along contour C_ϕ in the ϕ direction, we obtain

$$\oint_C \mathbf{H}_\phi \cdot d\mathbf{l} = H_\phi \cdot 2\pi r = NI = 0$$

because the contour C_ϕ along the circumferential direction encloses no current.

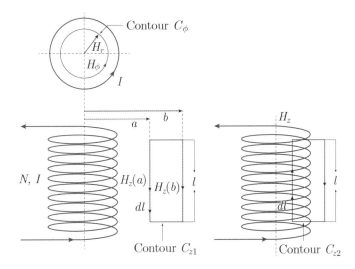

Figure 5.3: Magnetic flux generated in an N-turn long air-core solenoid.

From the above results, H_z is the only magnetic flux generated in the long air core solenoid. Applying Ampere's law along contour C_{z1}, we obtain

$$\oint_{C_{z1}} \mathbf{H}_z \cdot d\mathbf{l} = H_z(a)l - H_z(b)l = NI = 0$$

because contour C_ϕ encloses no current. Therefore, we obtain the relation $H_z(a) = H_z(b)$. When $b = \infty$, $H_z(b) = 0$. Hence, $H_z(a) = 0$, which indicates that there is no flux density outside the solenoid. Applying Ampere's law along contour C_{z2} while considering the magnetic intensity inside solenoid H_{in}, we obtain

$$\oint_{C_{z2}} \mathbf{H}_z \cdot d\mathbf{l} = H_{in}l = NI$$

Finally, we determine H_{in} as

$$H_{in} = \frac{NI}{l} \tag{5.12}$$

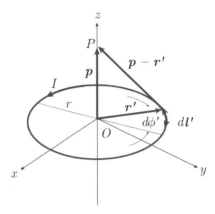

Figure 5.4: Magnetic flux intensity H_z at the center of a solenoid using the Biot-Savart law.

Instead of using Ampere's law, the magnetic flux intensity in an air-core solenoid having radius r can be obtained using the traditional approach, which is the Biot–Savart law. It states that at any point P, the magnitude of $\mathbf{H}$ produced by the differential element $d\mathbf{l}$ is proportional to current I, the length of $d\mathbf{l}$, and the sine of the angle lying between $d\mathbf{l}$ and a line connecting to P where the field is desired, as shown in Figure 5.4. It can be written concisely in a vector form as

$$\mathbf{H} = \oint_C \frac{Id\mathbf{l}' \times (\mathbf{p} - \mathbf{r}')}{4\pi|\mathbf{p} - \mathbf{r}'|^3} \tag{5.13}$$

where $\mathbf{p}$ and $\mathbf{r}'$ are position vectors of P and $d\mathbf{l}$, respectively, from center O. They are represented as

$$\begin{aligned} \mathbf{p} &= z\mathbf{a}_z \\ \mathbf{r}' &= r\mathbf{a}_r \\ d\mathbf{l}' &= rd\phi'\mathbf{a}_\phi \end{aligned}$$

where $\mathbf{a}_z$, $\mathbf{a}_r$, and $\mathbf{a}_\phi$ are unit vectors in the z, r , and ϕ directions. Then, we obtain

$$Id\mathbf{l}' \times (\mathbf{p} - \mathbf{r}') = Irzd\phi'\mathbf{a}_r + Ir^2d\phi'\mathbf{a}_z$$

Hence, $\mathbf{H}$ is represented as

$$
\begin{aligned}
\mathbf{H} &= \int_0^{2\pi} \frac{Irzd\phi'\mathbf{a}_r + Ir^2d\phi'\mathbf{a}_z}{4\pi(r^2 + z^2)^{3/2}} \\
&= \frac{Ir}{4\pi(r^2 + z^2)^{3/2}}[z\int_0^{2\pi}\mathbf{a}_r d\phi' + r\mathbf{a}_z\int_0^{2\pi}d\phi']
\end{aligned}
\tag{5.14}
$$

Since $\mathbf{a}_r = \cos\phi'\mathbf{a}_x + \sin\phi'\mathbf{a}_y$, the first part of Equation (5.14) is zero. Hence, we obtain

$$
\begin{aligned}
H_z &= \frac{Ir}{4\pi(r^2 + z^2)^{3/2}}(2\pi r) \\
&= \frac{Ir^2}{2(r^2 + z^2)^{3/2}}
\end{aligned}
\tag{5.15}
$$

When the one-turn solenoid is shifted from $z' = 0$ by $z = z'$ as shown in Figure 5.5(a), the magnetic field $\mathbf{H}$ generated due to the one-turn solenoid having the surface current density $\frac{NI}{l}$ is obtained by replacing I by $\frac{NI}{l}dz'$ from Equation (5.15) as

$$H_z = \frac{(NI/l)dz'r^2}{2(r^2 + (z - z')^2)^{3/2}}$$

As a N-turn solenoid is composed of the one-turn solenoid, as shown in Figure 5.5(b), magnetic field $\mathbf{H}$ due to the N-turn solenoid is obtained by integrating the magnetic field of the one-turn solenoid $z' = -\frac{l}{2}$ to $z' = \frac{l}{2}$ (note that $\int \frac{1}{(a^2+x^2)^{3/2}}dx = \frac{x}{a^2\sqrt{a^2+x^2}}$) as

$$
\begin{aligned}
\mathbf{H} &= \int_{-\frac{l}{2}}^{\frac{l}{2}} H_z \\
&= \int_{-\frac{l}{2}}^{\frac{l}{2}} \mathbf{a}_z \frac{(NIr^2/l)dz'}{2[r^2 + (z - z')^2]^{3/2}} \\
&= \mathbf{a}_z(\frac{NIr^2}{2l})\frac{-(z - z')}{r^2\sqrt{r^2 + (z - z')^2}}\Big|_{-l/2}^{l/2} \\
&= \mathbf{a}_z(\frac{-NI}{2l})[\frac{(z - l/2)}{\sqrt{r^2 + (z - l/2)^2}} - \frac{(z + l/2)}{\sqrt{r^2 + (z - l/2)^2}}]
\end{aligned}
\tag{5.16}
$$

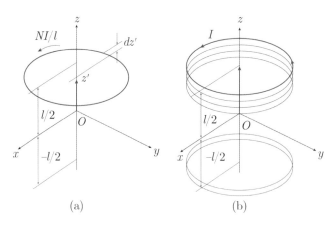

(a) (b)

Figure 5.5: (a) One-turn solenoid located at $z = z'$ and (b) N-turn solenoid composed of the one-turn solenoid.

From Equation (5.16), we obtain H_z at the center ($z = 0$) of a short-turn solenoid by assuming $l \cong 0$,

$$H_z(0) = \frac{NI}{2r} \tag{5.17}$$

For a long-turn solenoid, we obtain $H_z(0)$ by assuming $l \cong \infty$ as follows:

$$H_z(0) = \frac{NI}{l} \tag{5.18}$$

which is the same as Equation (5.12) that is obtained by applying Ampere's law. In general, the Biot–Savart law can be applied to analytically analyze magnetic systems. However, Maxwell's equations can also be applied for the same purpose because they provide an easier and more conceptual analysis of a magnetic system, though accurate analysis is not always feasible.

Figure 5.6 shows a magnetic system composed of iron and air cores. Geometrical parameters are graphically defined. In the iron cores, $B = \mu_o \mu_r H$ to represent a higher permeability of the iron core compared to that of the air core. μ_r is as high as 3,000. Now we use Maxwell's equations for magnetic system analysis. Equation (5.6) is applied along contour line C, which is one of the flux lines. Considering the flux intensities in the air cores, iron cores, and segment lengths of the contour line, we obtain

$$\oint_C \mathbf{H} \cdot d\mathbf{l} = H_1 l_1 + H_2 l_2 + H_3 l_3 + H_4 l_4 = NI \tag{5.19}$$

Suppose the magnetic flux flows through the same area, A, in the iron core without leakage. By applying Equation (5.9) on surface S, we obtain

$$\oint_S \mathbf{B} \cdot d\mathbf{A} = -B_1 A + B_3 A = 0 \qquad (5.20)$$

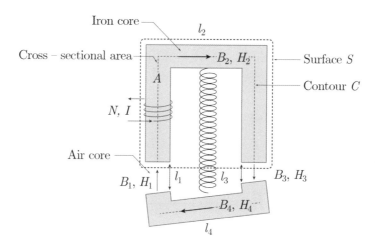

Figure 5.6: A magnetic system composed of iron and air cores.

One thing to note here is that H in the iron core should be negligibly small to justify that B represented by $\mu_o \mu_r H$ should exist. Hence, we can assume $H_2 = H_4 = 0$ because of the large μ_r. Based on these results,

$$NI = \frac{B_1 l_1}{\mu_o} + \frac{B_3 l_3}{\mu_o} = \frac{B_1 l_1 + B_3 l_3}{\mu_o} = \frac{B_1 (l_1 + l_3)}{\mu_o} \qquad (5.21)$$

Finally, B_1 is obtained as

$$B_1 = \frac{NI\mu_o}{(l_1 + l_3)} \qquad (5.22)$$

Then, the magnetic flux ϕ defined as $\int_S \mathbf{B} \cdot d\mathbf{A}$ is determined as

$$\phi = \frac{NI\mu_o A}{(l_1 + l_3)} \qquad (5.23)$$

Q1: Determine the flux densities in the air gaps in the magnetic system shown in Figures 5.7(a) and (b) when the current is I and the number of coil turns is N. The cross-sectional areas and lengths are graphically defined. Compare the fluxes in the center cores.

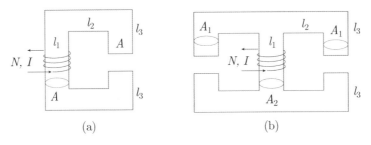

Figure 5.7: Magnetic systems having (a) single air and (b) double air gaps.

5.2.2 Electromagnetic forces

The force experienced by a test charge q moving with velocity $\mathbf{v}$ is [17]

$$\mathbf{F} = q\mathbf{E} + q\mathbf{v} \times \mathbf{B} \tag{5.24}$$

This is referred to as the Lorentz force [22]. The first term in Equation (5.24) is the force on a static charge. Since the moving charge is current, the second term is the force applied to the current. To consider the general cases of field systems, force density $\mathbf{f}(N/m^3)$ can be expressed as

$$\mathbf{f} = \rho_f \mathbf{E} + \mathbf{J}_f \times \mathbf{B} \tag{5.25}$$

where ρ_f and $\mathbf{J}_f$ are the charge and current densities having units of $(Coulomb/m^3)$ and (A/m^2), respectively. When only a magnetic field system is concerned, we have

$$\mathbf{f} = \mathbf{J}_f \times \mathbf{B} \tag{5.26}$$

Then, the average electromagnetic force, $\mathbf{F}$, is obtained by

$$\mathbf{F} = \int_V \mathbf{J}_f \times \mathbf{B} \; dV \tag{5.27}$$

The magnetic field system is actually used for producing the electromagnetic force which applies to mechanical systems. The electromagnetic force is generated at the air gap in the magnetic system whose magnetic flux was analyzed in Section 5.2.1. As an example, suppose that an N-turns coil is placed in the center of an air gap having length l_g, as shown in Figure 5.8. Here, d is the effective coil length occupied by B_g. We know the flux density, B_g, in the air

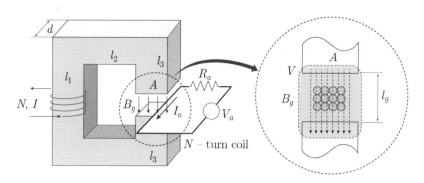

Figure 5.8: Electromagnetic force generated in an N-turn coil enclosed by the integral volume.

gap is calculated by referring to Equations (5.8) and (5.9) as follows:

$$B_g = \frac{NI\mu_o}{l_g}$$

When voltage V_a is supplied to the coil having resistor R_a, current I_a in the coil is $\frac{V_a}{R_a}$. Then, the electromagnetic force and its direction can be calculated using Equation (5.27) in consideration of integral volume V.

$$\mathbf{F} = \int_V \mathbf{J}_f \times \mathbf{B} \; dV = \int_V N\mathbf{I}_a \times \mathbf{B}_g \; dV = NB_g di_a = \frac{\mu_o dN^2 II_a}{l_g}(\rightarrow) \quad (5.28)$$

5.3   Magnetic circuit theory

As stated in Chapter 1, we can consider the three components of magnetic systems: inertia L, resistance R, and capacitor C like other physical systems, as listed in Table 1.1. Additionally, we can consider the following four variables: effort e, flow f, momentum p, and displacement q like other physical systems. However, in magnetic systems, there is no variable corresponding to momentum because the magnetic field is a kind of a wave. Hence, there is no component corresponding to the inertia.

The magnetic variables corresponding to effort e and flow f are the magneto-motive force M and magnetic flux rate $\dot{\phi}$. M is defined as

$$M = Ni \quad (5.29)$$

because current is considered to generate the magnetic field. Based on Equation (5.6), we obtain

$$M = Ni = \oint \mathbf{H} \cdot d\mathbf{l} \tag{5.30}$$

The magnetic variable corresponding to displacement q is magnetic flux ϕ because the displacement is obtained by integrating the flow. From the similarities among variables in physical systems, we obtain the following constitutive relations:

$$M = R\dot{\phi} \tag{5.31}$$

$$M = \frac{1}{C}\phi \tag{5.32}$$

Here, R is the magnetic resistance and C is the magnetic capacitance.

A magnetic field system can be equivalently considered an electrical system as it is constructed using an electric circuit composed of R and C, which are represented in Equations (5.31) and (5.32) if M and $\dot{\phi}$ correspond to voltage e and current i, respectively. Even though Equation (5.31) is a mathematically correct constitutive relation for R, the magnetic reluctance, $\Re$, is commonly defined using a different relation for the sake of enhancing conceptual understanding as

$$\Re \doteq \frac{M}{\phi} \tag{5.33}$$

For example, for an r-radius toroid core with current I supplied through N-turns, as shown in Figure 5.9, the effort and displacement variables of the magnetic system, M and ϕ, are expressed as

$$M = NI = \oint_C H dl = H \cdot 2\pi r \tag{5.34}$$

$$B = \mu_o \mu_r H \tag{5.35}$$

$$\phi = \int_A B dA \tag{5.36}$$

Then, the magnetic reluctance is obtained using its definition expressed in Equation (5.33)

$$\Re = \frac{M}{\phi} = \frac{NI}{BA} = \frac{2\pi r}{\mu_o \mu_r A} \tag{5.37}$$

Equation (5.37) indicates that the general formula of $\Re$ is expressed using the total length of the flux line, l, and cross-sectional area, A, as

$$\Re = \frac{l}{\mu_o \mu_r \cdot A} \qquad (5.38)$$

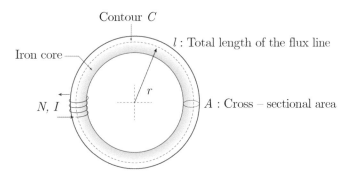

Figure 5.9: Toroidal core with supplied current.

The magnetic circuit can be represented using an electric circuit having $\Re$, as shown in Figure 5.10. The magneto-motive equation can be represented similarly to the voltage equation as

$$M = \Re \phi \qquad (5.39)$$

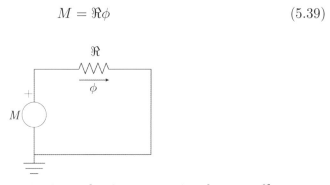

Figure 5.10: A magnetic circuit having magnetic reluctance $\Re$.

If there is an air gap in the toroid with gap length l_a, as shown in Fig 5.11(a), the equivalent magnetic circuit can be constructed, as shown in Figure 5.11(b). Here, $\Re_a$ and $\Re_c$ are the magnetic reluctances of the air gap and magnetic reluctance of the iron core, respectively. Then, the magneto-motive equation is

$$M = NI = \Re_a \phi_a + \Re_c \phi_c \qquad (5.40)$$

where ϕ_a and ϕ_c are the magnetic fluxes in the air gap and iron core, respectively.

ϕ_a and ϕ_c are the same because the reluctances are serially connected as currents in serially connected resistors are the same in an electric circuit. Thus, Equation (5.40) is arranged by referring to Equation (5.38) for expressing $\Re_a$ and $\Re_c$ as

$$M = \phi_a(\Re_a + \Re_c) = \phi_a\left(\frac{l_a}{\mu_o A} + \frac{l_c}{\mu_o \mu_r A}\right) \qquad (5.41)$$

ϕ_a is determined from Equation (5.41) when the design parameters are known. Compared to the case when there is only an iron core, ϕ_a is much reduced when there is an air core because $\Re_a$ is a large number.

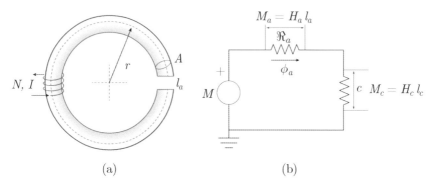

(a)   (b)

Figure 5.11: (a) A toroid with an air gap and length l_a and (b) a magnetic circuit having $\Re_c$ and $\Re_a$.

Q2: The second term of Equation (5.41) was ignored earlier to derive the magnetic flux density. Compare the effect of the air core in reducing the magnetic flux density when $r = 5$ cm, $A = 2$ cm^2, $N = 500$, $\mu_r = 2,000$, $I = 1$ A, $l_a = 1$ cm.

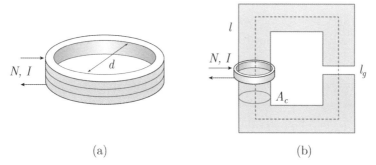

(a)   (b)

Figure 5.12: Illustrations of short-turn (a) air and (b) iron-core solenoids.

Owing to the magnetic circuit theory, we can predict that the short-turn air-core solenoid shown in Figure 5.12(a) has a low magnetic flux compared to the iron core solenoid shown in Figure 5.12(b). This is because the short-turn air core solenoid has a large magnetic reluctance. Let us investigate how small flux density the air core has compared to the iron core. The magnetic flux density of short turn solenoid, B_s at its center is obtained from Equation (5.17) as

$$B_s = \frac{\mu_0 NI}{2r}$$

If $N = 1000, I = 1.5\ A, d = 175\ mm$, $B_s = 1.077 \times 10^{-2}\ T$. A much higher flux density can be produced when the air core is replaced by a ferromagnetic material having the same diameter as the iron core. If the iron core solenoid has $l = 700\ mm$, $l_g = 4\ mm$, and $\mu_r = 1450$, as shown in Figure 5.12(b), the magnetic reluctances of the iron, $\Re_c$ and air gap, $\Re_a$ are respectively calculated from Equation (5.38) using the geometrical information as

$$\Re_c = \frac{l_c}{\mu_0 \mu_r A_c} = \frac{700 \times 10^{-3}}{(4\pi \times 10^{-7} \times 1450) \times (100^2 \times 10^{-6})} = 38.42 \times 10^3\ (A/Wb)$$

$$\Re_a = \frac{l_a}{\mu_0 A_a} = \frac{4 \times 10^{-3}}{(4\pi \times 10^{-7}) \times (100^2 \times 10^{-6})} = 318.3 \times 10^3\ (A/Wb)$$

Thus, the total equivalent reluctance of flux path $\Re_e$ is

$$\Re_e = \Re_c + \Re_a = 356.7 \times 10^3\ (A/Wb)$$

Then, the magnetic flux ϕ and density B are

$$\phi = \frac{M}{\Re_e} = \frac{1.5 \times 10^3}{356.7 \times 10^3} = 4.205 \times 10^{-3}\ Wb$$

$$B = \frac{\phi}{A} = \frac{4.205 \times 10^{-3}}{100^2 \times 10^{-6}} = 0.4205\ T$$

As a result, the ratio of flux density in the air gap to that produced in the center of the solenoid in the absence of iron is determined as

$$Ratio = \frac{0.4205}{1.077 \times 10^{-2}} = 39$$

We know that the magnetic flux in the iron core, ϕ_c, is the same as the magnetic flux in the air gap, ϕ_a, if there is no leakage flux. Let us investigate how the magnetic flux intensities in the core, H_c, and air gap, H_a, differ from each other.

$$H_c = \frac{B_c}{\mu_o \mu_r} = \frac{0.425}{1450 \times 4\pi 10^{-7}} = 230.8 \ (A/m)$$

$$H_a = \frac{B_a}{\mu_o} = \frac{0.425}{4\pi 10^{-7}} = 3.34 \times 10^5 \ (A/m)$$

It is interesting to note that $H_c \ll H_a$. If we rewrite Equation (5.40), we obtain

$$\begin{aligned} M &= M_c + M_a \\ &= H_c l_c + H_a l_a \end{aligned}$$

M_c is the magneto-motive force required to produce a flux density of $0.425 \ T$ in the core. M_a is the magneto-motive force required to produce the same flux density in the air gap. Using the above results and geometrical information we obtain $M_c = 161.5 \ A$ and $M_a = 1340 \ A$. Hence, we understand that a magneto-motive force, M_a, greater than M_c is required to produce the flux density in the air gap. The $M - \phi$ relationship for the iron core and air gap are shown in Figures 5.13(a) and (b), respectively. The hysteresis curve in $M_c - \phi$ is due to the magnetic property of the iron material [23]. The $M - \phi$ relationship for the complete magnetic system is obtained by adding the values of M_c and M_a for all values of ϕ, as shown in Figure 5.13(c). It is easily seen that ϕ is dominantly determined by M_a.

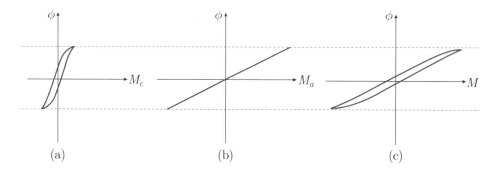

(a)　　　　　　　　　(b)　　　　　　　　　(c)

Figure 5.13: $M - \phi$ relationships for (a) the iron core and (b) air gap, and (c) the complete magnetic system.

5.4 Magnetic field analysis of a permanent magnet

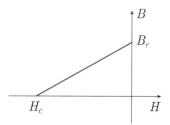

Figure 5.14: $B - H$ relationship of a permanent magnet.

A permanent magnet is modeled using the $B - H$ relationship, as shown in Fig 5.14. The relationship is mathematically represented as

$$B = B_r + \mu_o H \tag{5.42}$$

Because a permanent magnet is a type of material that is treated as a magnetic source, there is residual flux, B_r though there is no applied current, i.e., $B = B_r$ when $H = 0$. B_r is called the magnetic remanence. $-H_c$ is the magnetic coercivity which makes $B = 0$. M_c can be considered the demagnetizing intensity of the permanent magnet.

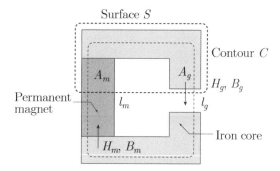

Figure 5.15: Magnetic flux in the air gap when there is a permanent magnet in the magnetic circuit.

Let us investigate how the magnetic flux in an air gap can be calculated when there is a permanent magnet and iron core in the magnetic circuit, as shown in Figure 5.15. Here, H_m, B_m, and A_m are the magnetic intensity, flux density, and area of the permanent magnet, respectively. H_g, B_g, and A_g are the magnetic intensity, flux density, and area of the air gap, respectively.

By applying the integral forms of the Maxwell equations to the contour C and surface S, we obtain

$$H_m l_m + H_g l_g \;=\; 0 \tag{5.43}$$

$$-B_m A_m + B_g A_g \;=\; 0 \tag{5.44}$$

Based on Equation (5.42), we obtain the following additional relation,

$$B_m = B_r + \mu_o H_m \tag{5.45}$$

When we assume $A_m = A_g$, we obtain $B_m = B_g$ from Equation (5.44). Using Equations (5.43) and (5.45), B_g is obtained as

$$B_g = \frac{B_r l_m}{l_m + l_g} \tag{5.46}$$

From Equation (5.46), it is concluded that $B_g \cong B_r$ if $l_g \cong 0$, which is as expected, because having no air gap does not reduce the magnetic flux generated by the permanent magnet. If l_m is large, $B_g \cong B_r$. If l_m is reduced, $B_g = \frac{B_r l_m}{l_g}$. These results indicate that the permanent length should be longer than the air gap length so that the magnetic flux is not reduced.

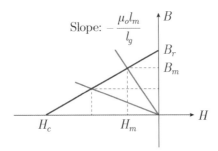

Figure 5.16: Graphical understanding of the B_m-H_m relationship when a magnetic load increases.

The magnetic flux generated by the permanent magnet can be graphically understood using the $B - H$ model curve represented by Equation (5.42). The slope of $\frac{B_m}{H_m}$ is obtained using Equations (5.43), (5.44), and (5.45) as

$$\frac{B_m}{H_m} = -\frac{\mu_o l_m}{l_g}$$

Figure 5.16 shows that B_m is decreased from B_r as l_g increases and l_m decreases because it has a lower slope. This indicates that a large l_g increases the magnetic load and causes much demagnetize the permanent magnet.

5.5 Magnetic field in a moving media

Now it is time to study the last equation described in Equation (5.10). We first need to study what $\oint_C \mathbf{E} \cdot d\mathbf{l}$ means. When voltage is supplied to the parallel plate capacitor shown in Figure 5.17, voltage V_{ab} is the potential difference between points a and b, which is defined as the work done for moving a positive unit charge from the bottom to the top [17].

$$V_{ab} = -\int_b^a \mathbf{E} \cdot d\mathbf{l} \tag{5.47}$$

Equation (5.47) can be rewritten again using the defined notations as

$$V_{ab} = \int_a^b \mathbf{E} \cdot d\mathbf{l} = El \tag{5.48}$$

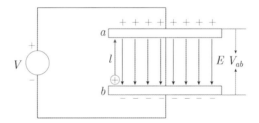

Figure 5.17: Definition of voltage V_{ab}.

The definition of voltage as the potential difference is similarly found in potential energy in the mechanical system. When there is a gravitational field $\mathbf{g}$ as shown in Figure 5.18, the potential energy E_p required for raising mass m to height h is

$$E_p = -\int_0^h m\mathbf{g} \cdot d\mathbf{l} = mgh \tag{5.49}$$

Additionally, the right part of Equation (5.10) shows how B or A changes with respect to time. Hence, we can say that voltage is induced when B or A

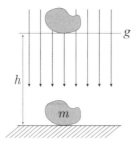

Figure 5.18: The definition of the mechanical potential E_p.

changes with respect to time. Figure 5.19 shows one example of showing how voltage V_{ab} is produced when magnetic field B changes with time inside the coil.

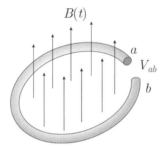

Figure 5.19: Voltage V_{ab} produced when magnetic field B changes with time only inside the coil.

By applying Equation (5.10), we obtain

$$\oint_C \mathbf{E} \cdot d\mathbf{l} = -A\frac{dB}{dt} \tag{5.50}$$

The left term of Equation (5.50) is

$$\oint_C \mathbf{E} \cdot d\mathbf{l} = \int_a^b \mathbf{E} \cdot d\mathbf{l} + \int_b^a \mathbf{E} \cdot d\mathbf{l} = \int_b^a \mathbf{E} \cdot d\mathbf{l} = -V_{ab} \tag{5.51}$$

Here, $\int_a^b \mathbf{E} \cdot d\mathbf{l} = 0$ because E in the conductor is zero. Thus, using Equations (5.50) and (5.51), we have

$$V_{ab} = A\frac{dB}{dt} \tag{5.52}$$

For an N-turn coil,

$$V_{ab} = NA\frac{dB}{dt} = \frac{Nd\phi}{dt} \tag{5.53}$$

$N\phi$ can be expressed using the flux linkage, λ, which is introduced in Table 1.1. λ is defined as the momentum variable of electrical systems. It can be understood by its name, which indicates how much magnetic flux ϕ is linked between adjacent coils. For example, when two coils are arranged as shown in Figure 5.20(a), $\lambda = 0$ because the values of ϕ produced in the two coils are orthogonal to each other. However, $\lambda = 2\phi$ in the two coils shown in Figure 5.20(b), because the values of ϕ are added together. Hence, Equation (5.53) can be rewritten as

$$V_{ab} = NA\frac{dB}{dt} = \frac{d\lambda}{dt} \tag{5.54}$$

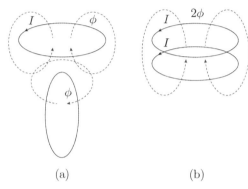

(a)   (b)

Figure 5.20: Illustration of (a) no magnetic flux in coils arranged orthogonal and (b) added magnetic flux in coils arranged in parallel.

As another effect that induces a voltage, time-varying area, $A(t)$ is also considered. Though magnetic flux B is constant, we can consider the area A that is enclosed by the contour C and changes with respect to time, as shown in Figure 5.21. Here, x is the position of the coil, m is the coil mass, and l is the effective length of the coil to which B is applied. As the varying area A expressed as $l \cdot x$, when the coil moves from an initial position x_o, the induced voltage, V_{ind} is

$$V_{ind} = \frac{d\lambda}{dt} = \frac{d}{dt}(NBA) = NBl\frac{dx}{dt} = NBlv$$

V_{ind} is usually referred to as the back electromotive force (EMF) to differentiate it from the applied voltage. From now on, it is denoted as e_b. Then, the induced current, I_{ind}, is calculated by considering the conductor resistor R as follows:

$$I_{ind} = \frac{V_{ind}}{R} = \frac{NBlv}{R}$$

I_{ind} flows in the direction where the induced magnetic flux produced by I_{ind} reduces the applied magnetic flux. This is called Rentz's law [22].

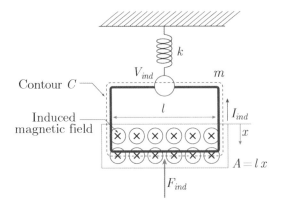

Figure 5.21: Voltage V_{ind} induced when the area A enclosed by C varies.

The induced force, F_{ind}, due to I_{ind} is calculated using Equation (5.27) as

$$F_{ind} = \int_V (\mathbf{J}_f \times \mathbf{B}\,)dV = \int (\frac{I_{ind}}{A_c}B)A_c dl = I_{ind}Bl = \frac{NBlv}{R}Bl = \frac{NB^2l^2v}{R} \tag{5.55}$$

where A_c is the area of the coil and dl is an infinitesimally small length of l. F_{ind} is found to be related to the velocity of the moving coil when Equation (5.55) is simply represented by $F_{ind} = cv$. Here, $c = \frac{NB^2l^2}{R}$. Referring to the constitutive relation introduced in Section 1.3.1, F_{ind} is known to be the damping force.

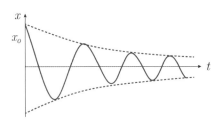

Figure 5.22: Motion of the second-order differential equation.

From the above analysis, we can obtain the mechanical system equation associated with the induced force as

$$-F_{ind} - kx = m\ddot{x} \tag{5.56}$$

Thus, we finally obtain

$$m\ddot{x} + \frac{B^2 l^2}{R}\dot{x} + kx = 0, \quad x(0) = x_o \tag{5.57}$$

Motion behavior of x is characterized by the natural frequency, ω_n, and the damping ratio, ζ, of the second-order differential equation, as shown in Figure 5.22. Refer to Section 4.3 for the time-response behavior of x.

Example 1: Figure 5.23 shows an electromechanical system composed of an iron core and a moving mass attached to a spring [17]. Current is supplied to the N-turn coil. The dimensions of the mechanical structure are graphically denoted. When the core in the center is in motion, we are interested in deriving the corresponding electrical equation.

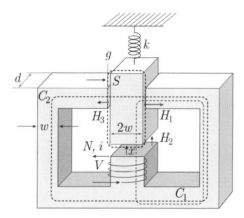

Figure 5.23: An electromechanical system composed of an iron core and a moving mass attached to a spring.

Solution 1: Applying Maxwell's equations to the line contours C_1 and C_2 and surface S, we obtain

$$
\begin{aligned}
H_1 g + H_2 x &= Ni \\
H_1 g - H_3 g &= 0 \\
\mu_o H_1 (wd) + \mu_o H_3 (wd) - \mu_o H_2 (2wd) &= 0
\end{aligned}
$$

From the above relations, we obtain

$$H_1 = H_2 = \frac{Ni}{g + x} \tag{5.58}$$

Thus, the magnetic flux ϕ through the air gap is

$$\phi = \int_S B_2 \cdot d\mathbf{A} = \frac{2wd\mu_o Ni}{g + x} \tag{5.59}$$

Flux linkage λ by the wire of N-turns is

$$\lambda = N\phi = \frac{2wd\mu_o N^2 i}{g + x} = L(x)i \tag{5.60}$$

where

$$L(x) = \frac{2wd\mu_o N^2 i}{g + x} \tag{5.61}$$

Then, input voltage V is described using Equation (5.53) as

$$\begin{aligned}
V &= Ri + \frac{d\lambda}{dt} \\
&= Ri + L(x)\frac{di}{dt} + i\frac{dL(x)}{dx}\frac{dx}{dt} \\
&= Ri + \frac{2wd\mu_o N^2 i}{g + x}\frac{di}{dt} - \frac{2wd\mu_o N^2 i}{(g + x)^2}v
\end{aligned}$$

where the last term is the back electromotive force (EMF), e_b

$$e_b = -\frac{2wd\mu_o N^2 i}{(g + x)^2}v \tag{5.62}$$

5.6 Bond graph representation of an electromechanical system

We can also represent an electromechanical system using a bond graph as it is composed of electrical, magnetic, and mechanical components. Because each component is a power system, there must be energy-storing elements $C's$ and energy-dissipating elements $R's$ with input effort or flow sources. The bond graph representation of an electromechanical system graphically shows how the electrical, magnetic, and mechanical components are related to each other.

First, we build up a bond graph for a magnetic system that has only C elements. The magnetic circuit theory discussed in Section 5.3 helps us to intuitively construct a bond graph with ease. For the magnetic system shown in Figure 5.7(a), the bond graph is represented in Figure 5.24 by assuming that there is no flux leakage outside the core. If voltage V is applied to the electromagnet, the electrical resistance is modeled using the R element. Voltage V and current i are converted to magnetic flux rate $\dot{\phi}$ and magneto-motive force M using a G-port because they have a cross relation to each other, i.e., $V = N\frac{d\phi}{dt}$ and $M = Ni$. 1-junctions represent the same flux rate, $\dot{\phi}$, across the iron cores and air gap.

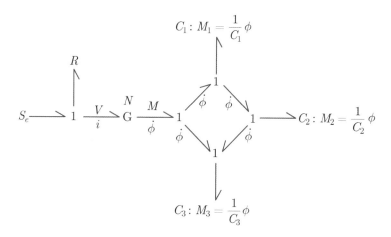

Figure 5.24: Bond graph representation of the magnetic system shown in Figure 5.7(a).

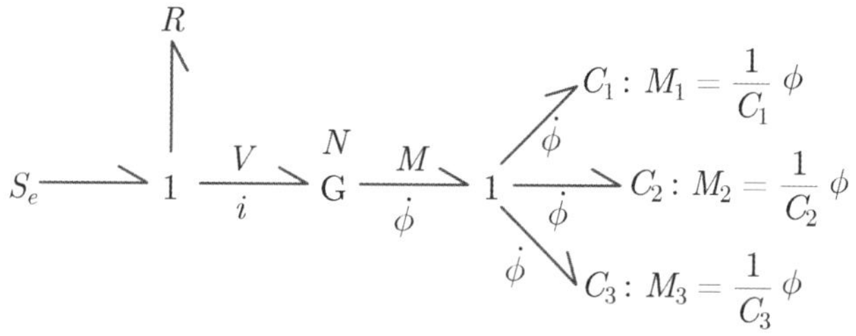

Figure 5.25: Simple representation of the bond graph shown in Figure 5.24.

M_1, M_2, and M_3 are the magnetic flux intensities of the upper core, air gap, and bottom core, respectively. C_1, C_2, and C_3 are the magnetic capacitances of the upper core, air gap, and bottom core, respectively. The bond graph shown in Figure 5.24 is represented again in Figure 5.25 based on the junction simplification rules. Moreover, the serial connection of the three C elements can be equivalently represented using effective capacitance C_{eff}.

Q3: Construct a bond graph for the same magnetic system shown in

Figure 5.7(a) when there are flux leakages in the upper and bottom cores.

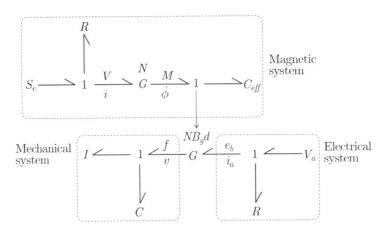

Figure 5.26: Bond graph of an electromechanical system (motor) that combines electrical, magnetic, and mechanical systems.

How about representing a bond graph when there is a moving conductor driven by force generated in the magnetic field? To achieve this representation, we revisit the electromechanical system shown in Figure 5.8. This system is a multidisciplinary system that combines electrical, magnetic, and mechanical systems. Referring to the electrical and mechanical equations associated with the electromechanical system, its bond graph is represented in Figure 5.26. Here, N is the number of coil turns, B_g is the magnetic flux density generated in the air gap, and d is the effective length of the coil occupied by B_g. The current in coil I_a generates mechanical force F with the relation $f = (NB_gd) \cdot I_a$. Its velocity v is obtained in return and produces e_b with the relation $e_b = (NB_gd) \cdot v$. Hence, the relation is represented using a G-port. (NB_gd) is considered as a modulus which relates the electrical system to the mechanical system and it is represented by the full arrow ($\rightarrow$) to show it is a signal flow, not a power or energy flow. f and v are delivered to inertial and spring elements.

The most popular electromechanical system is an electrical motor, which can be easily found in industrial applications. A direct-current (DC) motor, consisting of a stator and rotor, can be represented using the bond graph shown in Figure 5.26. The role of the stator is magnetic field generation, which is why it is commonly called a field [24]. The rotor is commonly called an amateur and its role is to supply current to a coil to produce the force, which is applied to a mass or spring in a mechanical system.

Unlike a motor, where force and speed are generated by current flowing through a conductor in a magnetic field, a mechanical system can also be braked by the induced current which is generated when force or speed is applied to a conductor in a magnetic field as shown in Figure 5.21. The bond graph corresponding to this case is shown in Figure 5.27.

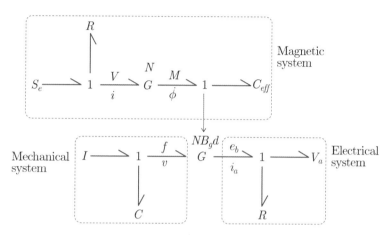

Figure 5.27: Bond graph of an electromechanical system (brake) that combines electrical, magnetic, and mechanical systems.

5.7 Various electromechanical systems

Many electromechanical systems have been developed to generate an electromagnetic force using a permanent magnet and electromagnet. The most famous example can be found in an acoustic speaker, in which the current in the electromagnet produces sound by compressing air. Owing to its feature of sound production, the acoustic speaker is conventionally called a voice coil motor (VCM) [25]. Several examples of recent applications can be found in [26],[27],[28],[29].

Figures 5.28(a) and (b) show a voice coil motor (VCM) nano scanner used in an atomic microscope (AFM) produced by EM4SYS Co., Ltd. [30]. The system is composed of a moving mass (plate), permanent magnets, and electromagnets. A flexure hinge structure is constructed to provide lateral stiffness of the scanner. The applied force, F_m, is obtained from the electromechanical principle in

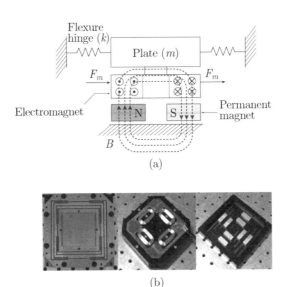

(a)

(b)

Figure 5.28: (a) Schematic diagram and (b) real configuration of a voice coil motor (VCM) nano scanner.

the magnetic system. The electromagnetic force, F_m, can be calculated using Equation (5.8) as

$$F_m = NBIl$$

where N is the number of coil turns, B is the flux density generated by the permanent magnets, I is the applied current to the electromagnet, and l is the effective length of the permanent magnet overlapped with the electromagnet. The plate of the VCM nano scanner moves back and forth by the applied current, I. Placing the electromagnets on the N and S permanent magnetic poles, F_m can be increased by a factor of two.

Instead of wired coils (electromagnet), conductive materials, such as copper and aluminum, can be used without winding, which causes a tethered difficulty in current supply. Of course, the induced current can be easily calculated in wired coils because it flows in a determined current path. However, when a copper disk is rotating in the magnetic flux generated in the pole, as shown in Figure 5.29, current is induced on the surface and the continuum media due to the time-varying area. This current is called an eddy current because it is generated in the form of a vortex around the pole area. The eddy current is induced in the direction in which it reduces the applied magnetic flux. If the disk rotates, the magnetic flux inside contour C is reduced as it rotates from

Position 1 to Position 2. Hence, torque is generated in the direction in which the speed of the rotating disk is reduced. The induced torque is modeled as a damping or braking torque because the eddy current is similar in nature to the induced current in a wired coil.

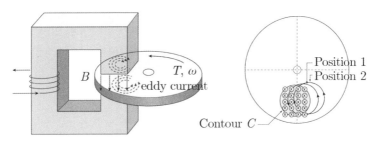

Figure 5.29: A copper disk rotating in magnetic flux generated in the core.

The in-depth torque-velocity relation of the rotating disk shown in Figure 5.29 can be found in literature [31],[32]. The braking torque, T_b, is simply expressed as

$$T_b = T_1 B^2 \omega = c\omega \tag{5.63}$$

where T_1 and B are the braking torque constant and magnetic flux density in the pole, respectively, and $c = T_1 B^2$.

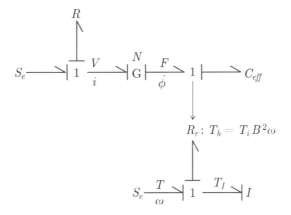

Figure 5.30: Bond graph model of an electromechanical system in which the eddy current generates a damping torque.

Considering that the torque–angular velocity relation is modeled using a resistance R_r element, we represent the bond graph model of the electromechanical system in Figure 5.30 for reference. Causality marks are applied to each bond to show how the effort and flow interact with each other. The causality mark on the R_r element shows that T_b is produced in return for applied angular velocity ω. It also shows that torque T_I, obtained by subtracting T_b from applied torque T, is acting on the inertia of the rotating disk.

This phenomenon of the eddy current is widely used for braking control in industrial applications, such as heavy-duty transportation vehicles, and the gyro drop in entertainment parks, as shown in Figure 5.31.

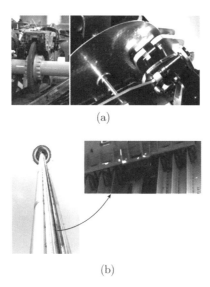

Figure 5.31: Non-contact brakes used in (a) a heavy-duty transportation vehicle and (b) a gyro drop in an entertainment park.

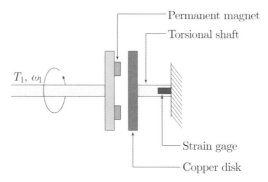

Figure 5.32: The non-contact braking torque sensing mechanism.

This non-contact braking torque is also used for torque sensing applications.

Force and torque are conventionally measured in a contact way. However, if we construct a simple mechanism composed of permanent magnets, a copper disk, a torsional shaft, and a strain gage, as shown in Figure 5.32, we can measure applied torque T_1 in a non-contact way using the strain gauge mounted in the stationary torsional shaft. This is because T_1 is the same as the braking torque, T_b. For electromechanical analysis, the bond graph model is represented in Figure 5.33, where the input angular velocity is ω_1. When the stationary copper disk is rotating slightly with ω_2 because of the induced torque, the braking torque, T_b is generated. Here, T_b is proportional to the relative angular velocity $\Delta\omega$ obtained by $\omega_1 - \omega_2$. Here ω_2 is the angular velocity of the copper disk. As a result of causality, T_b is determined to be the same as T_2 and it is transmitted to the copper disk (I element) and torsional shaft (C element). Here, T_I and T_3 are

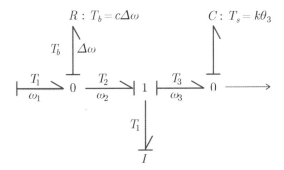

Figure 5.33: Bond graph of the non-contact braking torque sensing mechanism.

the torques delivered to the copper inertia and the torsional shaft, respectively. ω_3 is the angular velocity of the torsional shaft, θ_3 is the angle of the torsional shaft deformed by T_3. k is the stiffness of the torsional shaft. Based on the bond graph, we obtain the following torque relation:

$$T_1 = T_b = T_2 = T_I + T_3 = T_I + k\theta_3 \tag{5.64}$$

Thus, T_1 is directly measured by θ_3 when T_I is made to be small. θ_3 is measured using a strain gauge and corresponding electronic circuit. This measured signal, θ_3, is represented in the bond graph by the full arrow ($\rightarrow$) to show it is a signal flow.

5.8 Appendix

An infinitesimally small change in temperature, dT, can be expressed using a theorem of the partial derivative [20] as

$$dT = (\frac{\partial T}{\partial x})dx + (\frac{\partial T}{\partial y})dy + (\frac{\partial T}{\partial z})dz \tag{5.65}$$

This rule tells us how T changes when we alter all three variables, $dx, dy,$, and dz by infinitesimal amounts. Equation (5.65) can be rewritten in a more convenient form as

$$\begin{aligned}
dT &= (\frac{\partial T}{\partial x}\mathbf{i} + \frac{\partial T}{\partial y}\mathbf{j} + \frac{\partial T}{\partial z}\mathbf{k}) \cdot (dx\mathbf{i} + dy\mathbf{j} + dz\mathbf{k}) \\
&= \nabla T \cdot d\mathbf{l}
\end{aligned}$$

where $d\mathbf{l} = dx\mathbf{i} + dy\mathbf{j} + dz\mathbf{k}$ and ∇T is defined as

$$\nabla T = \frac{\partial T}{\partial x}\mathbf{i} + \frac{\partial T}{\partial y}\mathbf{j} + \frac{\partial T}{\partial z}\mathbf{k}$$

∇T is called the gradient of T. Hence, it can be interpreted that ∇T points in the direction of the maximum increase of function T. The gradient has the formal appearance of a vector ∇ multiplied by a scalar T.

$$\nabla T = (\frac{\partial}{\partial x}\mathbf{i} + \frac{\partial}{\partial y}\mathbf{j} + \frac{\partial}{\partial z}\mathbf{k})T \tag{5.66}$$

The divergence of $\mathbf{V}$ is mathematically defined using the ∇ operator as follows [20]:

$$\begin{aligned}
\nabla \cdot \mathbf{V} &= (\frac{\partial}{\partial x}\mathbf{i} + \frac{\partial}{\partial y}\mathbf{j} + \frac{\partial}{\partial z}\mathbf{k}) \cdot (V_x\mathbf{i} + V_y\mathbf{j} + V_z\mathbf{k}) \\
&= \frac{\partial V_x}{\partial x} + \frac{\partial V_y}{\partial y} + \frac{\partial V_z}{\partial z}
\end{aligned} \tag{5.67}$$

The divergence is a measure of how much vector V spreads out (divergence) from the point in question. The vector function $\mathbf{V}_1 = \mathbf{r}$ shown in Figure 5.34(a) has a large divergence at point P because it is spreading out. However, vector function $\mathbf{V}_2 = \mathbf{k}$ shown in Figure 5.34(b) has zero divergence at point P because it is not spreading out.

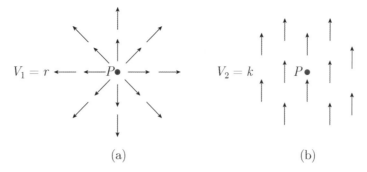

(a)   (b)

Figure 5.34: (a) Vector $\mathbf{V}_1$ having divergence and (b) vector $\mathbf{V}_2$ having no divergence.

The divergence of V_1 is obtained by applying Equation (5.67) as follows:

$$
\begin{aligned}
\mathbf{V}_1 &= \mathbf{r} = x\mathbf{i} + y\mathbf{j} + z\mathbf{k} \\
\nabla \cdot \mathbf{V}_1 &= \frac{\partial}{\partial x}(x) + \frac{\partial}{\partial y}(y) + \frac{\partial}{\partial z}(z) = 3
\end{aligned}
\tag{5.68}
$$

The divergence of V_2 is obtained by applying Equation (5.67) as follows:

$$
\begin{aligned}
\mathbf{V}_2 &= \mathbf{k} \\
\nabla \cdot \mathbf{V}_2 &= \frac{\partial}{\partial x}(0) + \frac{\partial}{\partial y}(0) + \frac{\partial}{\partial z}(1) = 0
\end{aligned}
\tag{5.69}
$$

The curl of $\mathbf{V}$ is mathematically defined using the ∇ operator as follows [20]:

$$
\begin{aligned}
\nabla \times \mathbf{V} &= (\frac{\partial}{\partial x}\mathbf{i} + \frac{\partial}{\partial y}\mathbf{j} + \frac{\partial}{\partial z}\mathbf{k}) \times (V_x\mathbf{i} + V_y\mathbf{j} + V_z\mathbf{k}) \\
&= (\frac{\partial V_z}{\partial y} - \frac{\partial V_y}{\partial z})\mathbf{i} + (\frac{\partial V_x}{\partial z} - \frac{\partial V_z}{\partial x})\mathbf{j} + (\frac{\partial V_y}{\partial x} - \frac{\partial V_x}{\partial y})\mathbf{k}
\end{aligned}
\tag{5.70}
$$

The curl measures a measure of how much vector $\mathbf{V}$ curls around the point in question. The vectors $\mathbf{V}_1$ and $\mathbf{V}_2$ shown in Figure 5.34 have no curl by physical interpretation or mathematical calculation results. How about vector $\mathbf{V}_3$ shown in Figure 5.35? As $\mathbf{V}_3$ is represented by

$$
\mathbf{V}_3 = -y\mathbf{i} + x\mathbf{j}
\tag{5.71}
$$

The curl of $\mathbf{V}_3$ is obtained by applying Equation (5.70) as

$$
\nabla \times \mathbf{V}_3 = (\frac{\partial}{\partial x}(x) - \frac{\partial}{\partial y}(-y))\mathbf{k} = 2\mathbf{k}
\tag{5.72}
$$

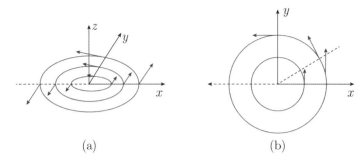

(a)   (b)

Figure 5.35: (a) Vector $\mathbf{V}_1$ having divergence and (b) Vector $\mathbf{V}_2$ having no divergence.

Control

—

Control

6.1   Why is a feedback loop needed?

It is best to make a good plant design that can perform tasks without an additional compensator, which is usually called a controller or filter. Therefore, we need to put all of our efforts into having good plant characteristics in terms of speed and accuracy. If an input is applied to a plant, the output response is decided based on its mathematical model. We can predict the output motion from the known model. However, a real system, which operates at high frequency, can be different from a mathematical model.

Figures 6.1(a) and (b) show the frequency responses of a real second-order mechanical system in terms of magnitude ratio and phase delay that differs greatly from a typical mechanical system composed of a rigid mass, damper, and spring that is represented by Equation (4.34). Sub-resonances, in addition to a fundamental resonance and large phase delay, are present. These differences often come from rigid body motion in the multiple degree-of-freedom axes of $x, y,$ and z axes and the rotational motion of those axes, as shown in Figure 6.2(a) [33]. Moreover, a rigid body can be modeled as a flexible body composed of several mass–damper–spring systems, especially at high frequency, which provides flexible body motion in multiple degree-of-freedom axes, as shown in Fig 6.2(b) [34]. These kinds of structure uncertainties or wrong identification of system parameters require a feedback loop to reduce the error between input and output.

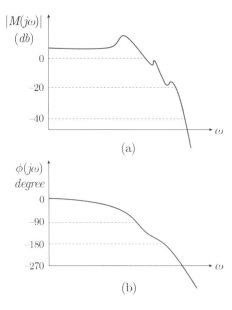

(a)

(b)

Figure 6.1: Frequency responses of a real second-order mechanical system, which differs from a typical mechanical system composed of a rigid mass, damper, and spring in terms of (a) magnitude ratio and (b) phase delay.

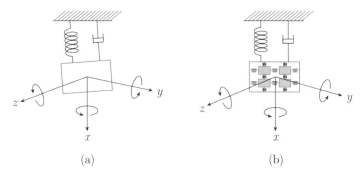

(a)   (b)

Figure 6.2: (a) Rigid body motion in multiple degree-of-freedom x, y, and z axes and rotational motion about those axes and (b) flexible body motion composed of several mass–damper–spring systems.

For example, when a unit input is applied to an object that is modeled as a mass, spring, and damper system, we know the displacement of the object would be in a steady state if the parameters were the same as those of the mathematical model. However, the displacement of the object is different from the result obtained from the mathematical model because there are differences in modeling between the parameters in reality.

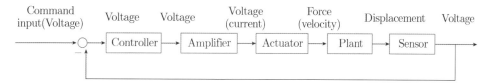

Figure 6.3: Block diagram of a position feedback system.

Figure 6.3 shows a block diagram of a typical example of a position feedback system composed of a controller, amplifier, actuator, plant, and sensor. The amplifier is made of electrical components and provides voltage or current from a power supply to the actuator. The actuator converts electrical variables, such as voltage or current, to mechanical power variables, such as force or velocity; the plant reacts according to a given force or velocity.

The sensor is used for measuring the position information to know if it is larger or smaller than the command input. Therefore, negative feedback is required for comparison between the input and output. Because position control needs to be realized, we need a displacement sensor, which provides a voltage that is proportional to its magnitude. Figures 6.4(a) and (b) show how the sensor output voltage varies for the position of, for example, mass x. When the sensor measures the positive $(+x)$ and negative $(-x)$ positions, as shown in Figure 6.4(a), it provides positive $(+V)$ and negative $(-V)$ voltages, respectively. In this case, the command input is zero for moving the mass to the center position.

However, in the case where the sensor can measure only the positive position, as shown in Figure 6.4(b), the command input is V_c for moving the mass to the center position, x_c. When the mass moves in the positive direction $(+x)$ from the center position, the sensor signal is larger than the command input, which produces a negative error signal (command input signal-sensor output signal). Then, the actuator supplies a negative force to the mass so that it moves closer to the center position. Finally, the sensor signal can be the same as the command input, which results in zero error. When the sensor signal is smaller than the command input because of the negative directed movement, the positive error supplies a positive force to the mass so that it moves close to the center position again. In conclusion, though the plant is not modeled well in terms of parameter estimation, system structure or the system particularly has modeling uncertainties, the sensor output reflects the current situation. Therefore, the error signal can compensate the plant so that it can follow the command input.

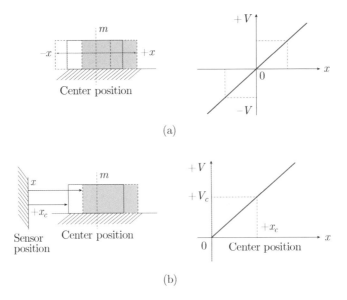

Figure 6.4: Sensor output voltage differed from the position of a mass (a) when the positive $(+x)$ and negative $(-x)$ positions are measured and (b) when only the positive position is measured.

A plant may be designed not to realize its performance as expected. For example, it can have a low have a low damping ratio that causes large vibrations in motion. It can have a low bandwidth, which causes a large magnitude reduction and phase delay in an output signal at a high frequency. A controller can be added in front of the amplifier, as shown in Figure 6.3 to modify the plant and improve plant performance in terms of speed and accuracy. Sometimes, the plant can be unstable, which yields to serious damage to the plant. This instability problem can also be solved using a controller.

There are two analysis methods in the controller design process: closed-loop analysis (time-response analysis) and open-loop analysis (frequency-response analysis). First, we begin with closed-loop analysis. It has the advantage of being a straightforward process and can be easily understood in a feedback system. However, this method is very complicated and tedious, requiring the repetition of a job whenever the order of a plant is increased and the controller changes. In this analysis, the controller cannot be separated from the plant. Next, we will show the advantages of open-loop analysis, which include a simple controller design process that modifies the frequency characteristics of the plant.

6.2　Feed-forward transfer function

The feed-forward transfer function, $G(s)$, includes the amplifier, actuator, plant, and sensor. It is very important not to make the amplifier and sensor cause additional dynamic effects on the plant because they can affect the plant dynamics in terms of magnitude and phase delay. If possible, they need to be chosen or designed so that their bandwidths are higher than that of the plant. $G(s)$ is also called an open-loop transfer function. Because they can affect the plant dynamics in terms of magnitude and phase delay. If possible, they need to be chosen or designed so that their bandwidths should be higher than that of the plant.

If the dynamic characteristics of the amplifier and sensor are negligibly small, $G(s)$ for a second-order system can be represented as

$$G(j\omega) = K_a K_t K_s \frac{\omega_n^2}{\omega_n^2 - \omega^2 + j(2\zeta\frac{\omega}{\omega_n})} \tag{6.1}$$

where K_a, K_t, and K_s are the gains of the amplifier, actuator, and sensor, respectively. The corresponding Bode plot is shown in Figure 6.5. As addressed in Section 4.4.5, it has a peak magnitude $\omega_r = \omega_n\sqrt{1 - 2\zeta^2}$, $M_r = |G(j)|_{\max} = \frac{K_a K_t K_s}{2\zeta\sqrt{1-\zeta^2}}$.

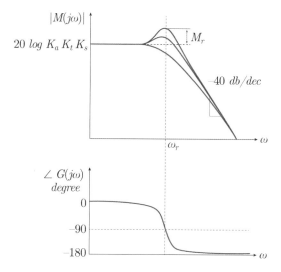

Figure 6.5: Feed-forward transfer function, $G(s)$, with minimal amplifier and sensor dynamics when the plant is a second-order mechanical system.

6.2.1 Voltage amplified loop

There are two types of amplifiers—a voltage amplifier and a current amplifier—for driving an actuator. Depending on the amplifier used, there are different models for control configuration. If a voltage amplifier is used for driving an electrical motor only having inertia J, the mechanical equation associated with the motor is

$$T = J\frac{d\omega}{dt} \tag{6.2}$$

Here, T is the torque produced by the motor, and ω is the angular velocity of the motor. The electrical equation associated with the motor is obtained as

$$V = L\frac{di}{dt} + Ri + e_b \tag{6.3}$$

Here, L and R are the inductance and resistance, respectively, of the coil in the electric motor. V is the amplified voltage, i is the current applied to the motor, and e_b is the back electromotive force (EMF) induced when the angular velocity of the motor, ω, is generated in the magnetic field, as addressed in Section 5.5. As there is a moving conductor (rotating coil) in the magnetic field, this electromechanical system can be understood by the electromechanical principle and is represented by the bond graph shown in Figure 5.26. Hence, torque T generated by the motor and ω have the following relations:

$$T = K_t i \tag{6.4}$$
$$\omega = K_e e_b \tag{6.5}$$

where K_t and K_e are defined as the force and electromotive force constants, respectively.

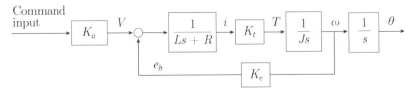

Figure 6.6: Block diagram of a voltage amplifier in an electric motor.

The electric motor with a voltage amplifier having gain K_a is alternatively represented using a block diagram in Figure 6.6. There is an internal angular velocity feedback loop in a motor when a voltage-amplified loop is used. Hence, ω is controlled to be constant by V. For example, when T is increased, ω will increase. Then, e_b, which is proportional to ω, also increases. As a result, the error $V - e_b$ will decrease, thereby decreasing i. Hence, the reduced current will finally regulate the value of ω to its original value before it increases.

What will happen when a robot is operated using a voltage amplifier? Because of the internal velocity feedback loop in the electric motor, as shown in Figure 6.6, the robot can rotate at a constant angular velocity that is proportional to the amplified voltage, V. However, if the end effector of the robot gets stuck to an obstacle that prevents it from rotating, the motor will continue to draw maximum current to make the robot rotate with an angular velocity proportional to V. Then, the motor will eventually burn out because the maximum current is supplied for a long period of time through a power supply. Therefore, it can be said that the voltage-amplified loop should not be used when it happens to contact other objects.

6.2.2   Current amplified loop

Figure 6.7 shows a block diagram of an electric motor in a robot operated using a current amplifier with gain K_a. The current i is amplified in proportion to the command input voltage, V, and generates constant torque which rotates the electric motor in the robot. The block diagram has no feedback loop. Therefore, there must be a steady-state error in the angle, unless K_a, K_t, and J are correctly modeled. When the end effector of the robot happens to be stuck to an obstacle, the current flows only as much as the given force. Therefore, it can be said that a current amplifier should be used when a robot happens to contact an object. However, the use of a current amplifier can cause a steady-state error if there are modeling uncertainties. To solve the problem, angle feedback control can be considered to reduce the steady-state error.

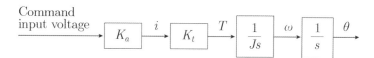

Figure 6.7: Block diagram of a current amplifier.

The block diagram shown in Figure 6.7 corresponds to a single link of the robot manipulator. When a robotic system with multi-link and multi-joint is considered, the structure of the control block diagram looks more complicated. However, the block diagram of a single link structure can be the fundamental concept for understanding a more complicated robotic system. It is necessary to understand the kinematic and dynamic equations of robotic systems for constructing the overall block diagram.

Example 1: Draw a control block diagram of a robotic system composed of a two-link and two-joint mechanism, as shown in Figure 6.8 for position control x, y of the end effector. Each joint is driven by an electrical motor using a current amplifier. M_1 and L_1 are the mass and length, respectively, of Link 1. M_2 and L_2 are the mass and length, respectively, of Link 2. θ_1 and θ_2 are the rotating angles of Links 1 and 2, respectively. There are encoders on the shafts of the two motors to measure the rotating angles.

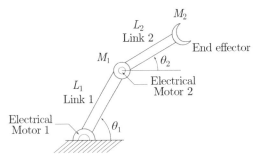

Figure 6.8: A robotic system composed of a two-link mechanism.

Solution 1: The governing equations of the two link mechanism are represented as follows [35],[36]:

$$T_1 = f_1(\ddot{\theta}_1, \dot{\theta}_2^2, \theta_1, \theta_2)$$
$$T_2 = f_2(\ddot{\theta}_2, \dot{\theta}_1^2, \theta_1, \theta_2)$$

where T_1 and T_2 are the applied torques generated by electrical motors 1 and 2, respectively. As indicated in the governing equations, the robotic system is a nonlinear and coupled system. If the kinematic relation between the rotating angles (θ_1, θ_2) and position (x, y) is expressed using nonlinear transformation matrix $[T]$, we obtain

$$\begin{bmatrix} x \\ y \end{bmatrix} = [T] \begin{bmatrix} \theta_1 \\ \theta_2 \end{bmatrix} \tag{6.6}$$

Equation (6.6) indicates that the desired position of the end effector (x, y) can be obtained in terms of desired angles θ_{r1} and θ_{r2} using the inverse matrix of $[T]$. Since θ_{r1} and θ_{r2} can be controlled using a feedback loop of angles θ_1 and θ_2 measured by sensors such as rotary encoders, we obtain $\theta_1 \cong \theta_{r1}$ and $\theta_2 \cong \theta_{r2}$.

Figure 6.9 shows a control block diagram of electric motors of the robotic system based on the above-mentioned dynamic and kinematic equations when a current amplifier is used. Here, V_{rx} and V_{ry} are the command input voltages corresponding to the desired position of the end effector (x, y). i_1 and i_2 are the current amplified and applied to the motor to produce the torques T_1 and T_2. $[T]^{-1}$ is used to calculate the desired angles θ_{r1} and θ_{r2} from the desired position of the end effector (x, y). K_s is a sensor gain

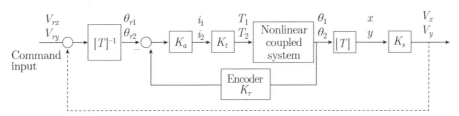

Figure 6.9: A robotic system composed of a two-link mechanism.

When the target position of the end effector (x, y) can be measured using a position sensor with gain K_a, it can be also fed back to the position command input to improve the position accuracy using the feedback loop denoted using the dashed line. However, a commercial sensor to measure its position is not easily available on the market. Hence, it is usual to just control the electrical motors using precise encoders with gain K_r that can be easily attached to the motor shaft. We can call the system shown in Figure 6.9 a closed-loop system for rotating angle control, but it can be said to be an open-loop system for position control if a proper position sensor is not installed. Therefore, it is not guaranteed that the target position is accurately controlled even though the links can rotate accurately because of the parameters uncertainties and dynamical effects associated with the actual links.

It is important to know which sensor is available on the market and is easier to implement for feedback control because the sensor selection can change the structure of the control block diagram. In addition, it can affect the system accuracy.

Q1: Draw the control block diagram of a drone system composed of motors and rotating blades as shown in Figure 6.10 when it is controlled to fly to a desired location in the air using a global positioning system (GPS), which can measure its absolute position. What are the differences between the robotic system shown in Figure 6.9 and the drone system in terms of control structure?

Figure 6.10: A drone system flying to a desired location in the air.

6.3 Analysis of the closed loop system

6.3.1 Proportional (P) controller

To improve system performances such as transient response characteristics and steady-state error, a controller can be added to the feed-forward loop. The system performances with a controller can be analyzed in the time domain.

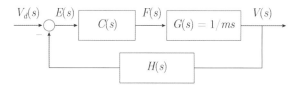

Figure 6.11: A feedback system represented using a block diagram and a proportional (P) controller.

When applied force $f(t)$ is applied to a mass with the velocity output, as shown in Figure 4.11, the standard form of a feedback system is represented in Figure 6.11 using a block diagram, where the open loop transfer function $G(s) =$

$1/ms$, feedback transfer function $H(s)$, and the controller transfer function is $C(s)$. The output velocity, v, is fed back to the desired input, v_d, so that velocity error e is decreased. For example, when v is bigger than v_d, negative force f is applied to reduce v. When v is smaller than v_d, positive force f is applied to increase v.

From Figure 6.11, we have the relations of

$$
\begin{aligned}
E(s) &= V_d(s) - V(s)H(s) \\
F(s) &= C(s)E(s) \\
V(s) &= G(s)F(s)
\end{aligned}
$$

where $V_d(s), E(s), F(s)$, and $V(s)$ are the Laplace transforms of reference $v_d(t)$, error $e(t)$, force $f(t)$, and output $v(t)$, respectively. Hence, closed loop transfer function G_{cl} defined as $\frac{V(s)}{V_d(s)}$ is obtained as

$$
\frac{V(s)}{V_d(s)} = \frac{C(s)G(s)}{1 + C(s)G(s)H(s)} \tag{6.7}
$$

Controller $C(s)$ is called a proportional (P) controller when the controller gain is constant K_p. Then, using $C(s)$, $G(s)$, and $H(s) = 1$, we obtain

$$
G_{cl} = \frac{K_p}{K_p + ms} = \frac{1}{1 + \frac{m}{K_p}s} \tag{6.8}
$$

What is the difference between $G(s)$ and G_{cl}? We know that the integrator represented in G changes to a low-pass filter in G_{cl}.

Q2: Sketch the time-domain response, v, of the open and closed loop systems when $v_d = \sin(10t)\ m/s$, $m = 0.5\ kg$, and $K_p = 10$ for the system shown in Figure 6.11. How should the proportional controller gain be adjusted to have a faster time response in the closed loop system?

Figure 6.12: A mechanical system composed of mass and damper.

What will happen when a damper is added to a mass in a mechanical system, as shown in Figure 6.12? Here, we are interested in the position. The dynamic equation corresponding to the mechanical system is $f = m\frac{dv}{dt} + cv, v = \frac{dx}{dt}$ thus, we obtain the open loop transfer function $G(s)$ with input force and output position expressed as

$$G(s) = \frac{X(s)}{F(s)} = \frac{1}{s(ms + c)} \tag{6.9}$$

As stated before, a drift motion problem, which could eventually collide toward one side, exists in this case due to the integrator. A feedback system using a P controller, $C(s)$, can be constructed to solve this problem, as shown in Figure 6.13.

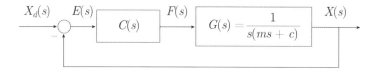

Figure 6.13: A proportional controller for a plant composed of mass and damper.

$X_d(s)$ and $X(s)$ are the Laplace transforms of reference $x_d(t)$ and output $x(t)$, respectively. Thus, we obtain

$$E(s) = X_d(s) - X(s) \tag{6.10}$$
$$F(s) = C(s)E(s) \tag{6.11}$$
$$X(s) = G(s)F(s) \tag{6.12}$$

Using Equations (6.10) through (6.12), the closed loop transfer function of the system, $G_{cl}(s) = \frac{X(s)}{X_d(s)}$ with $C(s) = K_p$, is represented as

$$G_{cl}(s) = \frac{C(s)G(s)}{1 + C(s)G(s)} = \frac{K_p}{ms^2 + cs + K_p} = \frac{\omega_n^2}{s^2 + 2\zeta\omega_n s + \omega_n^2} \tag{6.13}$$

where $\omega_n = \sqrt{\frac{K_p}{m}}$ and $\zeta = \frac{1}{2}\frac{c}{\sqrt{mK_p}}$. By comparing Equation (6.9) and Equation (6.13), we can recognize that the mass and damping system changes into a system having a mass, damper, and electrical spring with stiffness K_p. Graphically, a mass-damper system in the open loop shown in Figure 6.14(a) is changed to a mass-damper-electric spring system in the closed loop system, as shown in Figure 6.14(b).

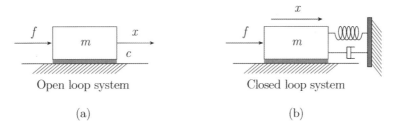

Figure 6.14: (a) A mass-damper system in an open loop system and (b) a mass-damper-electric spring system in the closed loop system.

One more thing to note here is that $x(t)$ is almost the same as $x_d(t)$ when K_p is increased further, as expected from Equation (6.13).

This result, $x(t) \approx x_d(t)$ is also investigated using an error analysis, which can be made using a transfer function of $\frac{E(s)}{X_d(s)}$. Using Equations (6.10) through (6.12) again, $\frac{E(s)}{X_d(s)}$ is obtained as

$$\frac{E(s)}{X_d(s)} = \frac{1}{1 + C(s)G(s)} = \frac{ms^2 + cs}{ms^2 + cs + K_p} \qquad (6.14)$$

For example, when a unit step input $1(t)$ is applied, the steady state error, $e(t) = x_d(t) - x(t)$ is obtained using the final value theorem [13] as follow:

$$\lim_{t \to \infty} e(t) = \lim_{s \to 0} sE(s) = \lim_{s \to 0} s \frac{ms^2 + cs}{ms^2 + cs + K_p} X_d(s) = \frac{ms^2 + cs}{ms^2 + cs + K_p} \cong 0$$
$$(6.15)$$

Here, the relation of $X_d(s) = \frac{1}{s}$ is used.

Q3: Investigate the steady state error of the feedback system shown in Figure 6.13 when ramp input t is applied.?

6.3.2   Proportional-derivative (PD) controller

The transfer function of a proportional and derivative (PD) controller can be represented by $K_p(1 + T_d s)E(s)$, where K_p and T_d are the proportional and derivative gains, respectively. Thus, the PD controller has error and error derivative outputs, i.e., $K_p(e + T_d \frac{de}{dt})$. A block diagram of a closed loop system for the plant transfer function $G(s) = \frac{1}{s(ms+c)}$ is represented in Figure 6.15.

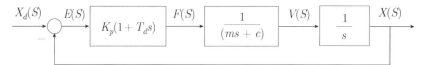

Figure 6.15: Block diagram of a closed loop system with a PD controller.

Closed-loop transfer function, G_{cl}, is then obtained as

$$G_{cl} = \frac{C(s)G(s)}{1 + C(s)G(s)H(s)} = \frac{K_p(1 + T_d s)}{ms^2 + (c + K_p T_d)s + K_p} \qquad (6.16)$$

Equation (6.16) indicates that the mass-damper system (first-order low-pass filter and integrator) is converted to a second-order low-pass filter system with a natural frequency due to K_p and an increased damping ratio due to K_p and T_d. Particularly, a mass-damper system in an open loop is changed to a mass-electric damper–electric spring in a closed-loop system. Therefore, a PD control feedback system can take advantage of an increasing damping ratio compared with the P control feedback system, as indicated in Equation (6.13).

As stated regarding the P and PD controllers, the advantage of the time domain approach is that it is a straightforward and mathematical analysis. The controller gains of $C(s)$ can be determined by investigating its output response behavior.

6.3.3   Stability

A plant is considered stable when the roots of the characteristic equations of the plant or system are located in the left half s plane [37]. The roots of the characteristic equation are called the poles of the system. By referring to Equation (4.43), we know that the poles of a second-order system, $s = \sigma \pm j\omega_d = -\zeta\omega_n \pm j\omega_d$. A positive value of s means a negative $\zeta\omega_n$, which places the pole in the right half plane and results in instability. This can be explained by referring to Equation (4.41) in which $e^{-\zeta\omega_n t}$ is equal to infinity as $t = \infty$.

For example, the system shown in Figure 6.12 has poles $s = 0$, $s = -\frac{c}{m}$, which are obtained from Equation (6.9). The system is marginally stable because the pole $s = 0$ is not located in the left half plane of s. However, when a PD controller is used, we obtain a new characteristic equation as follows:

$$ms^2 + (c + K_p T_d)s + K_p = 0 \qquad (6.17)$$

Then, the poles of the closed-loop system are obtained as

$$s = \sigma \pm j\omega_d = -\zeta\omega_n \pm j\omega_n\sqrt{1-\zeta^2} \qquad (6.18)$$

where $\omega_n = \sqrt{\frac{K_p}{m}}$ and $\zeta = \frac{1}{2}\frac{c+K_pT_d}{\sqrt{mK_p}}$. The new poles are depicted in the s plane shown in Figure 6.16 along with poles of the open loop system for comparison. The poles of the closed-loop system s_1 and s_2 are located far away from $s = 0$, which means higher stability is obtained in a closed loop system.

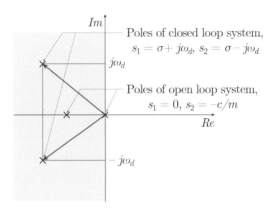

Figure 6.16: Poles of open loop and closed loop systems.

Roots s_1 and s_2 vary depending on K_p and T_d, as represented in Equation (6.18). This analysis is called root locus analysis [37]. We can use the s plane to determine K_p and T_d because they are related to ω_n and ζ which are represented by Eqns. (4.44) and (4.45), respectively. The closed loop system response due to the effects of ω_n and ζ can be analyzed in the time domain by referring to Section 4.3.

6.3.4 Physical meaning of P, PD, PI controllers

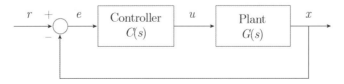

Figure 6.17: Standard form of a feedback loop with a controller.

Figure 6.17 shows the standard form of a feedback loop with a P controller. Here, r, e, u, and x are standard notations of the command input, error, controlled output, and output signal, respectively. Suppose a unit-step input of $r = 1$ is applied to a plant. When x is smaller than 1, the error, $e = r - x$, is positive, which generates a positive u because u is proportional to e. Thus, the increased u makes x larger. Eventually, e is zero in the case of $x = r$, which results in no controlled output and no production of motion. However, if there is an inertia element in the plant, it will not stop and x becomes larger than 1. In this case, a negative e is generated, which produces a negative controlled output in a similar manner. This again makes x smaller than r. Hence, this type of positive and negative-controlled output repeats until x is finally equal to r. Hence, the output response x has a large overshoot and takes time to reach the steady state value, as shown in Figure 6.18(a).

When a PD controller is used, the derivative signal, $\frac{de}{dt}$, is added to the proportional signal, e, whose signals are displayed using dashed and solid lines, respectively, in Figure 6.18(b). The PD controller effect creates a negative input (braking force) u at t_{PD} that is earlier than that with a P controller, which occurs at t_P, as shown in Figure 6.18(c). The negative u helps to reduce the plant speed with a small overshoot. Therefore, the PD controller can make the plant reach a steady state faster. These advantages are obtained by the leading phase created due to differentiation. However, this leading phase can increase the electronic noise because it amplifies the magnitude ratio at a higher frequency.

The transfer function of a proportional and integral (PI) controller can be represented by $K_p(1 + \frac{1}{T_i s})$ where K_p and T_i are the proportional and integral gains, respectively. Thus, the PI controller has error e and error integral outputs, i.e., $K_p(e + \frac{1}{T_i} \int e dt)$. When a PI controller is used, the integral signal $\int e dt$ is added to e whose signals are displayed using dashed and solid lines, respectively, in Figure 6.19(a). It is seen that the PI controller provides a positive input to the plant, even when a negative input is produced due to the P controller. As a result, the PI controller effect provides a negative input (braking force) u, that occurs at t_{PI}, which is later than that of the P controller, which occurs at t_P, as shown in Figure 6.19(b). This feature can make the plant unstable.

The advantage of using the PI controller can be found in an application, where there is friction associated with the motion of the plant. When controlled output u is smaller than the friction, we have a persistent steady-state error

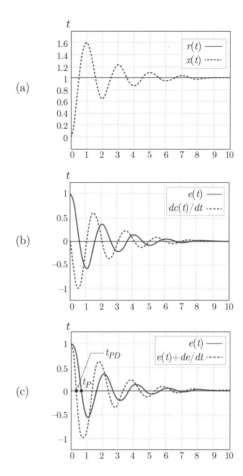

Figure 6.18: (a) Output response x when a P controller is used, (b) error e and error derivative $\frac{de}{dt}$ signals, and (c) the controlled output of a PD controller.

because the plant will not move if the applied force is smaller than the friction force. However, if this error is amplified by the PI control action, the amplified error can move the plant when it is larger than the friction force, reducing the steady-state error.

6.3.5   Closed loop representation with disturbances and noise

There exist other inputs in a closed-loop system in addition to the command input. Let us suppose a ball is moving on the floor due to the applying force f and it experiences friction force f_f and external vibration d, as shown in Figure 6.20. f_f and d are considered disturbances. The corresponding dynamic

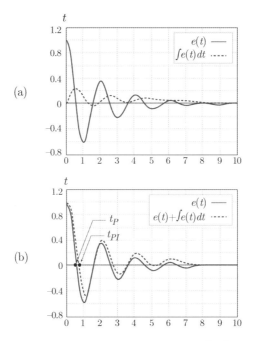

(a)

(b)

Figure 6.19: (a) The error e and error integral $K_p(e + \frac{1}{T_i} \int edt)$ signals and (b) the controlled output of a PI controller.

equation is expressed as

$$F - F_t = ms^2 X(s) \tag{6.19}$$

where F and F_f are the Laplace transforms of f and f_f, respectively.

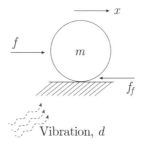

Figure 6.20: A moving ball experiencing a friction force and external vibration.

When position control is required to move the ball to a desired distance with the disturbances, f_f and d, they need to be included in a feedback system. According to Equation (6.19), f_f and d are considered additional inputs to the plant as shown in Figure 6.21. Suppose that the position controller is $C(s)$, the plant is $G(s)$ represented by $\frac{1}{ms^2}$, and the position sensor gain is K_s. Then, the

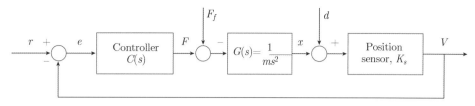

Figure 6.21: A feedback system with friction force and external vibration inputs.

output, V is obtained by the transfer function $\frac{V(s)}{F_f(s)}$ using the relation of

$$[((r - V)C - F_f)G + d]K_s = V(s)$$

Thus, $\frac{V(s)}{F_f(s)}$ is obtained with the assumption of $r(s) = 0$ and $d(s) = 0$, as

$$\frac{V(s)}{F_f(s)} = \frac{K_s G(s)}{K_s C(s)G(s) + 1} = \frac{K_s}{ms^2 + K_s C(s)} \tag{6.20}$$

Equation (6.20) describes how the output is affected by the friction force input. If $C(s)$ is increased $V(s)$ is less affected by the friction force disturbance.

When force control is considered to apply a desired force, its block diagram can be constructed similarly to the position control. For example, when a needle is approaching a rigid object as shown in Figure 6.22, the needle needs to be controlled to make contact with the object without breaking. The needle approaches the object until contact force f is the same as the desired force, f_d, using a force controller. A light and rigid needle can be modeled as a spring having coefficient k, where x is the displacement of the needle moving due to controlled output u.

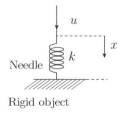

Figure 6.22: Schematic drawing of a needle for force control.

The block diagram associated with the force feedback system is shown in Figure 6.23. Suppose that x is proportional to u by actuator gain G_{act}. Because force output must be measured, a force sensor is used. Just after the needle makes contact with the object, it should withdraw as quickly as possible before it experiences a force larger than f_d so that it does not break.

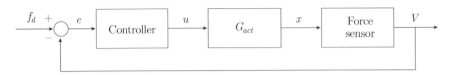

Figure 6.23: Block diagram associated with the force feedback system.

Figure 6.24 shows the time response of force control when low controller gains are used, such as K_p in the P and PD controller. However, when a faster response is required, we need to increase the controller gains to obtain a higher bandwidth. However, this can increase the overshoot of the output force due to the decreased damping ratio $\zeta = \frac{1}{2}\frac{c}{\sqrt{mK_p}}$. When the output force is larger than f_{max}, as indicated by the dotted line in Figure 6.24, the needle breaks. Hence, we need to take caution when controller gains are increased because it can cause system instability though it can increase system bandwidth. There are several force control methods available for securing stability and improved system performance in addition to conventional controllers [38],[39].

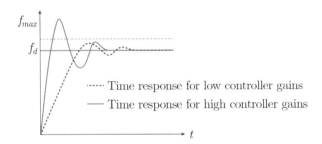

Figure 6.24: Time responses of force when a rigid and flexible needle is used for contact.

6.4 Analysis of an open loop system

6.4.1 P controller design for mechanical systems

The problem in the closed-loop analysis is that it is impossible to perform the analysis without knowing $C(s)$. It is required to know $C(s)$ and $G(s)$ simultaneously to predict the output response. However, it is more effective to design $C(s)$ after we know how $G(s)$ needs to be compensated for performance improvement in terms of stability, speed, and steady-state error.

When the open loop transfer function (feed-forward transfer function) is $C(s)G(s)$, as shown in Figure 6.17, the transfer function of the closed loop system, $G_{cl}(s)$, is

$$G_{cl}(s) = \frac{C(s)G(s)}{1 + C(s)G(s)} = \frac{G'(s)}{1 + G'(s)} \tag{6.21}$$

The transfer function of the closed-loop system in the frequency domain, $G_{cl}(j\omega)$, is

$$G_{cl}(j\omega) = \frac{G'(j\omega)}{1 + G'(j\omega)} \tag{6.22}$$

The closed loop output is unstable when $G'(j\omega) = -1$. This condition can be separated as $|G'(j\omega)| = 1$ and $\angle G'(j\omega) = -180°$. Hence, it can be said that the closed system is unstable when $|G'(j\omega)| = 1$ and $\angle G'(j\omega) = -180°$. Here, we define the gain margin as the reciprocal of the magnitude $|G(j\omega_{pc})|$ at phase crossover frequency ω_{pc}, where $\angle G(j\omega_{pc}) = -180°$. A phase margin is the amount of additional phase lag at gain crossover frequency ω_{gc} that is required to bring the system to the verge of instability.

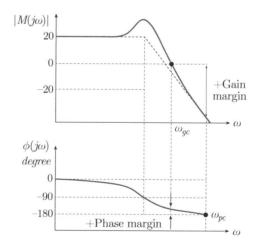

Figure 6.25: Positive gain and phase margins of the second order system $G(j\omega)$.

Figure 6.25 shows the phase and gain margins of an ideal second-order system $G(j\omega)$ composed of a mass, spring, and damper. This system has a positive phase margin determined at ω_{gc} and a positive gain margin determined at ω_{pc}, as represented by the Bode plot in Figure 6.25. It is important to note here that when P controller $C(j\omega)$ is used, open-loop function $G'(j\omega) = C(j\omega)G(j\omega)$ also has positive gain and phase margins because the phase of $G'(j\omega)$ does not decrease below $-180°$ even though the magnitude ratio of $G'(j\omega)$ increases. Hence, the closed-loop system is always stable. Thus, open loop analysis has the advantage that the closed loop stability can be investigated by just using the open loop transfer function, $C(j\omega)G(j\omega)$ without the need for closed-loop system analysis.

However, it happens that $G(j\omega)$ has a greater negative slope than $-40\,dB/decade$ in reality, which yields a bigger phase delay than $-180°$. In this case, if the P controller gain increases for better system performance, $G'(j\omega)$ may have negative gain or phase margins, which makes the closed-loop system unstable. Hence, controller gains should be selected appropriately. For example, suppose that $C(j\omega)G(j\omega)$ is represented using the solid line in Figure 6.26. It has a delay larger than $180°$ at a high frequency. However, it has a positive gain margin (GM) and a positive phase margin (PM). However, when the P controller gain is increased, $C(j\omega)G(j\omega)$ represented by the dotted line turns out to have a negative GM and a negative PM. Particularly, the high gain of the P controller makes the system unstable. Therefore, a lower P controller gain is required to obtain a positive PM.

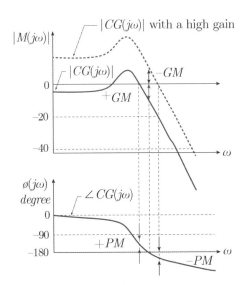

Figure 6.26: Gain and phase margins changed by increasing the P controller gain.

Q4: Most real-world secondary mechanical systems exhibit phase de-

lays greater than 180° at high frequencies, different from the frequency characteristics shown in Figure 6.25. Why does this happen?

6.4.2 PD controller design for mechanical systems

Let us investigate how the PD controller, $C(j\omega)$, affects the GM and PM compared with the case when a P controller is used for the same plant, $G(j\omega)$, as shown in Figure 6.26. Figure 6.27(a) shows $C(j\omega)G(j\omega)$ for the case when the PD controller has a lower cutoff frequency due to high T_d. The cutoff frequency of the PD controller should be designed so that derivative action should occur at an operating frequency range of interest. Compared with the P controller, the GM and PM are increased to ensure a stable feedback system because of the phase lead effect of the PD controller. However, when K_p increases, it can be unstable because the crossover frequency is shifted to a higher frequency where those margins are decreased. Similar results are also obtained as shown in Figure 6.27(b) when the PD controller has a higher cutoff frequency due to the low value of T_d.

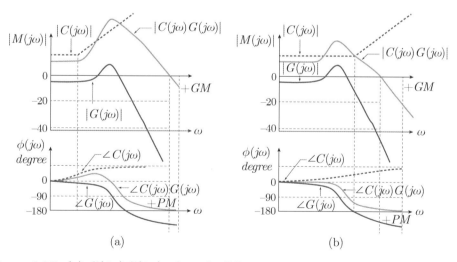

Figure 6.27: (a) $C(j\omega)G(j\omega)$ when the PD controller has a lower cutoff frequency and (b) $C(j\omega)G(j\omega)$ when the PD controller has a higher cutoff frequency.

Additionally, the closed-loop system performance can be determined by the

GM and PM of $G'(s) = C(j\omega)G(j\omega)$ as shown below [40], [41].

$$\omega_{bw} = \omega_{gc} \text{ when } PM \leq 90° \tag{6.23}$$

$$\omega_{bw} = 2\omega_{gc} \text{ when } PM \leq 45° \tag{6.24}$$

$$\zeta = \frac{PM}{100} \text{ when } PM \leq 70° \tag{6.25}$$

$$M_r = |G(j\omega)|_{max} = \frac{1}{2\zeta\sqrt{1-\zeta^2}} = \frac{1}{2\sin(\frac{PM}{2})} \tag{6.26}$$

Q5: When a closed loop system is implemented using a PD controller, we have $K_p(1 + T_d s)$ for the second-order mechanical plant composed of mass (m), spring (k), and damper (c), as shown in Figure 4.3. Here $m = 0.1\ kg$, $k = 360\ N/m$, and $c = 3.6\ Nsec/m$.

a) Design a PD controller so that the closed loop bandwidth is 80 rad/sec and the damping ratio is 0.7.

b) What is the steady state error at the low frequency when a unit step input is applied?

c) When a unit step input is applied, what is the output response?

d) Determine the PM and compare the expected result based on the control design rule and the bandwidth of the closed loop system, 80 rad/sec.

6.4.3 Dynamic effect of sensing system on control

From the above analysis, it can be concluded that the gains of the P, PD, and PI controllers cannot simply increase if $G(j\omega)$ is a higher-order system. In addition to $G(j\omega)$, there are other subsidiary components, such as power amplifiers and sensors, which can cause an increase in the phase delay of a closed-loop system. Therefore, it is essential to check the amplifier and sensor to verify if they provide additional phase delay effects on the plant as many electronic circuits are associated with these components.

For example, it is normal to design a sensor such that the noise at high frequency is reduced because sensors are used for accurately measuring the output signal. Hence, there is a signal processing circuit that uses electronic filters such as a low-pass filter, which causes phase delay. Therefore, it is very important to experimentally obtain the dynamic characteristics of the individual components using the frequency response technique before feedback control is implemented.

Furthermore, it is necessary to check if the bandwidths of these components are sufficiently high compared to that of the plant to avoid deterioration of the overall system performance. The sensor bandwidth is recommended to be at least 10 times higher than the system's operating bandwidth [42].

6.5   Compensator design

We can use PI and PD controllers to respectively decrease the steady state error and increase the PM of plant $G(j\omega)$. However, the integrator in the PI controller can deteriorate the stability because of its 90° phase delay. The differentiator in a PD controller can increase the electronic noise because of its amplification at a higher frequency. The weakness of PI and PD controllers can be improved using lag and lead compensators.

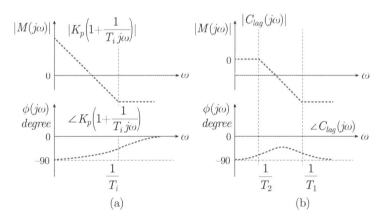

Figure 6.28: Magnitude ratios and phase delays of (a) a PI controller represented by $K_p(1 + \frac{1}{T_i j\omega})$ and (b) lag compensator $C_{lag}(j\omega)$.

A PI controller is modified by a lag compensator, which cuts off an integrating action at a lower frequency, to have a smaller phase delay. The lag compensator, $C_{lag}(j\omega)$, can be represented by

$$C_{lag}(j\omega) = \frac{T_1 j\omega + 1}{T_2 j\omega + 1}, \quad T_1 < T_2 \tag{6.27}$$

where T_1 and T_2 are the coefficients of the compensator. $\frac{1}{T_1}$ and $\frac{1}{T_2}$ are called the cut-off frequencies of the compensator. Figures 6.28(a) and (b) show the magnitude ratios and phase delays of a PI controller represented by $K_p(1+\frac{1}{T_i s})$ and lag compensator $C_{lag}(j\omega)$, represented by Equation (6.27), respectively for a comparison purpose. As seen in Figure 6.28(b), the range of integral action is reduced with reduced phase delay. T_1 and T_2 should be selected so that the integral action occurs at the operating frequency range of interest.

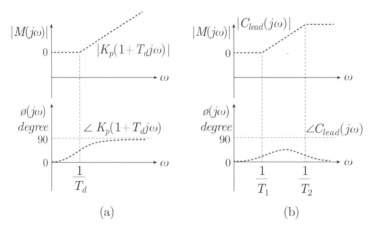

Figure 6.29: Magnitude ratios and phase delays of (a) a PD controller and (b) lead compensator $C_{lead}(j\omega)$.

A PD controller can be modified by a lead compensator, which cuts off a differentiating action at a higher frequency to reduce signal amplification. The lead compensator can be similarly represented by Equation (6.27), where $T_1 > T_2$. Figures 6.29(a) and (b) show the corresponding magnitude ratios and phase delays of PD controller, $K_p(1 + T_d j\omega)$ and lead compensator $C_{lead}(j\omega)$, respectively, for comparison purposes. As seen in Fig 6.29(b), the range of derivative action is reduced while the signal amplification is reduced at a higher frequency. T_1 and T_2 should be selected so that the derivative action occurs at the operating frequency range of interest.

A PID (proportional-integral-derivative) controller is constructed by combining PI and PD controllers. A lag–lead compensator is also constructed by combining lag and lead controllers. The magnitude ratios and phase delays of a PID controller and lag–lead compensator are shown in Figures 6.30(a) and (b), respectively. By applying the phase lag and lead characteristics of each controller $C(j\omega)$ to plant $G(j\omega)$, the gain and phase margins of $C(j\omega)G(j\omega)$ are properly adjusted.

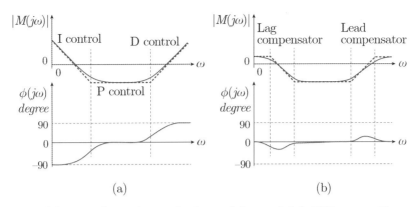

Figure 6.30: Magnitude ratios and phase delays of (a) PID controller and (b) lag-lead compensator.

Example 3: Consider the system shown in Figure 6.31. When you draw a Bode plot, the system has sufficient phase (PM) and gain margins (GM); thus, the controller gain has room to increase even though the PM decreases due to the increasing gain. It is important to design a controller that provides sufficient PM as well as a high gain to decrease steady-state error. Let us suppose that the controller gain is first set to 10 to provide a small steady-state error. Then, from $10\,G(jw) = 40/(jw(jw + 2))$, as shown in Figure 6.32, the phase and gain margins of the system are found to be 18° and $+\infty$ dB, respectively. A PM of 18° implies that the closed loop system is quite oscillatory. Thus, the specification calls for a PM of at least 50°. Hence, we need to find a controller to provide an additional phase lead of 32°. To achieve this goal, a lead compensator is chosen and designed.

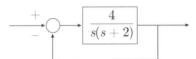

Figure 6.31: Control system $G(s) = 4/(s(s + 2))$.

Solution 2: The trial-and-error method is first conducted through some procedures to determine cutoff frequencies while considering the lead angle. Thus, the transfer function of lead compensator $C(s)$ is designed with the determined gain and the additional phase as follows:

$$C(s) = 10\,\frac{0.2s + 1}{0.05s + 1} \qquad (6.28)$$

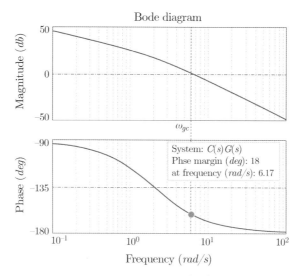

Figure 6.32: Bode diagram of $10\, G(jw) = 40/(jw(jw+2))$.

Figure 6.33 shows the results of $C(s)G(s)$ with a PM of approximately $50°$ obtained at $\omega_{gc} \approx 10\ rad/sec$ and the expected high GM.

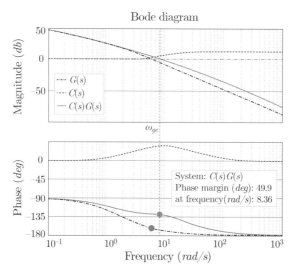

Figure 6.33: Bode diagram of $C(jw)G(jw)$.

To check the design rule-of-thumb suggested in Equations (6.24) and (6.25), the frequency response of closed-loop transfer function $G_{cl}(s)$ is obtained as

$$G_{cl}(s) = \frac{C(s)G(s)}{1 + C(s)G(s)} = \frac{40(0.2s + 1)}{0.05s^3 + 1.1s^2 + 10s + 40} \qquad (6.29)$$

The Bode plot of $G_{cl}(s)$ is shown in Figure 6.34. The closed-loop bandwidth is about 10 rad/sec, which is almost the same as w_{gc} depicted in Figure 6.33.

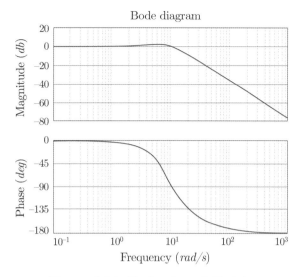

Figure 6.34: Bode plot of $G_{cl}(s)$.

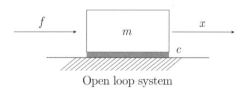

Open loop system

Figure 6.35: A mechanical plant composed of mass m and damper c.

Q6: Figure 6.35 shows a mechanical plant composed of mass, m and damper, c, in which force F is applied to the mass to produce x. Here, $m = 1\ kgr$, $c = 20\ \pi N sec/m$, and open-loop gain K is $72,000\ N$. Assume that π equal 3. The position sensor has a phase lag described as

$$\frac{100}{s + 100}$$

a) Draw the frequency response of the output to input x/F, which is converted to voltage in a Bode plot.

b) When $\sin(100t)$ is applied to the plant as an input voltage, sketch the output position measured in the sensor and compare it with the input voltage in the time domain.

c) When a closed system is implemented with P controller gain $K_p = 1$, is it stable or unstable, and why? Determine the phase and gain

margins and indicate them in a Bode plot.

6.6 Design of control components for good command following

Figure 6.36 shows a real configuration for position control of a mechanical system. Here, V_d, e, u, i, F, x and V_x are the command input voltage, error, controlled output, current, force, position, and sensor voltage, respectively, corresponding to the position; K_a, K_t, and K_s are the amplifier gain, force constant, and sensor gain, respectively. The unit of the command input, V_x, is the voltage corresponding to output x that is determined by K_s. Here, K_a and K_s are assumed to be constant.

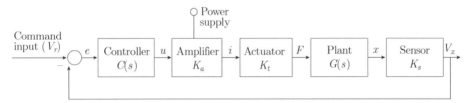

Figure 6.36: A real configuration for position control of a mechanical system.

Q7: What is the command input of the closed loop system shown in Figure 6.36 when a position sensor is used?

As seen in Figure 6.36, there are several components in the closed loop system whose characteristics are represented by $C(s)$, K_a, K_t, $G(s)$, and K_s. The closed-loop transfer $\frac{V_x(s)}{V_r(s)}$ is

$$\frac{V_x(s)}{V_r(s)} = \frac{G'(s)}{1 + G'(s)}$$

where $G'(s) = K_a K_t K_s G(s)$. For good command following, we should have the condition $V_x(s) \cong V_r(s)$, which can be realized if $G'(s)$ has a large magnitude. Ideally, if $G'(s)$ is infinity, $V_x(s) = V_r(s)$. Then, does increasing K_a and K_s help $G'(s)$ increase and improve command following performance?

Suppose we have $C(s) = 1$, $K_a = 1\,(A/V)$, and $K_s = 1\,(V/m)$ for a system having $K_t = 1\,(N/A)$, $G(s) = 1\,(m/N)$ as the first case. Let us compare this to the second case, $K_a = 10\,(A/V)$ and $K_t = 0.1\,(\frac{N}{A})$ when other gains are not changed, as well as the third case, $K_a = 10$, $G(s) = 0.1 \, (m/V)$, where other gains are also not changed. We obtain the same results for the open loop transfer functions of the three cases, i.e., $G'(s) = 1$. Do they have the same performance? If they do not, which case exhibits a better performance?

In the simulation, there are no differences in the results of the three cases in terms of system performance, such as stability and bandwidth, because they look like they have the same $G'(s)$. This is a weakness of the simulation. In real systems, however, it is not possible to increase K_a in the second and third cases because a large current can burn out the coil in the actuator. Therefore, the current is constrained from increasing more than the maximum current, i_{max}, which also determines the maximum force. Particularly, we need to consider voltage or current constraints in controller design. These types of electrical effects cannot be easily recognized in simulation. Of course, it is recommended to use a high power supply, if allowed, to provide sufficient voltage or current as long as they do not damage any actuator parts. Additionally, the high supply current caused by high controller gains in $C(s)$ should be considered.

Instead of increasing K_a, it is more important to design an actuator and plant that provide high gains of K_t or $G(s)$, as the example in the first case, because they have the same effect of supplying a smaller current to the plant. Particularly, because it has room for increasing current, a higher force can be exerted, which results in better system performance where $V_x(s) = V_r(s)$.

Moreover, it is also necessary to check whether all control components shown in Figure 6.36 linearly operate for the input signal. If voltage and current are not sufficiently supplied, we will not obtain the expected control performance. For example, Figure 6.37 shows the output signal $x(t)$ of the closed loop system of a plant when different step input magnitudes are applied. When voltage and current are sufficiently supplied from the current and voltage amplifiers to the actuator, we obtain similar output response shapes for increased inputs, as shown in Figure 6.37. These results are expected because the plant is a linear system. However, what will happen when the unit step input is increased further? It is expected that the voltage and current will eventually run out and no longer be linear due to insufficient power supply or actuator malfunction. As a result, we do not have linear output responses. Therefore, it is important to check whether the controller generates linear signals for increased inputs. When nonlinear signals due to voltage or current saturation are generated in any control components, a closed loop system is not capable of providing a

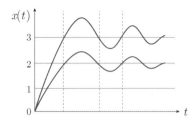

Figure 6.37: Output responses of a linear closed loop system when different magnitudes of step inputs are applied.

satisfactory result in output response.

Now, it is time to investigate, from a control point of view, how the applied voltage and current affect the motion of a mechanical plant (system). As mentioned in Section 6.2.1, the current and back-emf (EMF) voltage generated in an electromechanical system are proportional to its force and velocity, respectively. Hence, we need to provide sufficient voltage and current to the actuator to generate the force or velocity required for a mechanical plant.

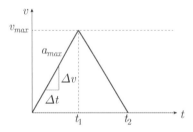

Figure 6.38: Trapezoidal motion profile.

Suppose we have a trapezoidal motion profile [43] for a mechanical plant represented in the time domain, as shown in Figure 6.38. since the current in an electromechanical actuator is proportional to the force generated, the applied maximum current i_{max} determines the maximum acceleration, a_{max}, i.e., the slope of the motion profile. Therefore, it is recommended to adjust the control gain to generate as much current as possible. If a maximum current is exerted, optimal control is realized. According to Equation (6.3), the applied maximum voltage determines the maximum velocity, v_{max}, because the EMF voltage produced in an electromechanical actuator is proportional to the velocity.

Trapezoidal motion profile $v(t)$ should be generated to make the output follow the input well. Particularly, the frequency of $v(t)$ needs to be within its control bandwidth. Otherwise, higher-speed input is meaningless because the plant is not capable of following the input. Suppose that the output for the ramp input is obtained with a small time delay using velocity control, as shown in Figure 6.39(a) with a dotted line, which indicates that this input frequency is a control bandwidth. The distance traveled with this profile is assumed to be S. How does the output velocity respond for inputs whose time profiles are higher and lower than the bandwidth, as shown in Figures 6.39(b) and (c)? The motion profiles shown in Figures 6.39(b) and (c) correspond to shorter and longer distance movements, respectively.

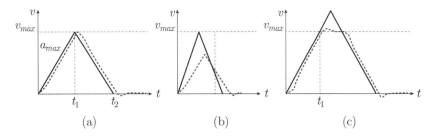

Figure 6.39: Responses for different trapezoidal profiles.

Output responses for these two cases are shown with dotted lines in Figures 6.39(b) and (c). Because the frequency of input $v(t)$ in Figure 6.39(b) is higher than the control bandwidth, the output response is delayed. Because the frequency of input $v(t)$ in Figure 6.39(c) is lower than the control bandwidth, the output response follows the input well. However, the output response is saturated over time t_1 because $v(t)$ is larger than v_{max}. Hence, when a longer distance must be traveled, a longer period of a trapezoidal input is required.

Q8: What type of input should be applied when a distance shorter than S is required?

6.7   Stability of an electronic system

Figure 6.40(a) shows an electronic feedback system constructed using an operational amplifier. v_{in} and v_o are the input and output voltages, respectively, i_{in} is the current through input resistor R_1, and i_f is the current through feedback resistor R_f. The closed-loop transfer function of the operational amplifier can be obtained using a block diagram shown in Figure (7.5), where the command input is v^+, the output is v_o, and the plant is $G(s)$. The error signal to the operational amplifier is $v^+ - v^-$. Here, v^- is obtained by the voltage division rule of v_{in} and v_o, i.e.,

$$v^- = \frac{R_f}{R_1 + R_f}v_{in} + \frac{R_1}{R_1 + R_f}v_o \tag{6.30}$$

Voltage $v_{in}(s)$ can be viewed as a type of disturbance from the control point of view. Then, $v_{in}(s)$ is obtained from the algebraic relations between the components of $v_o(s)$ plant $G'(s)$ of the operational amplifier and the feedback loop gain $H(s)$ defined by $\frac{v^-(s)}{v_o(s)}$ as

$$v^+(s) - \left(\frac{R_f}{R_1 + R_f}v_{in}(s) + \frac{R_1}{R_1 + R_f}v_o(s)\right)G(s) = v_o(s) \tag{6.31}$$

Since $v^+(s) = 0$ and $G(s)$ is assumed to be very large, $\frac{v_o(s)}{v_{in}(s)}$ is simply expressed from Equation (6.31) as

$$\frac{v_o(s)}{v_{in}(s)} = -\frac{R_f}{R_1}, \tag{6.32}$$

The closed-loop transfer function $\frac{v_o(s)}{v^+(s)}$ is obtained by assuming disturbance $v_{in} = 0$ and $G(s) = \infty$ as

$$\frac{v_o(s)}{v^+(s)} = \frac{G(s)}{1 + G(s)H(s)} \cong \frac{1}{H(s)} = \frac{R_1 + R_f}{R_1} \tag{6.33}$$

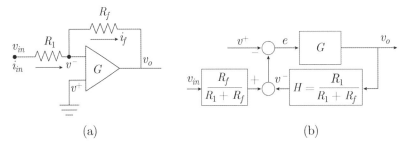

(a) (b)

Figure 6.40: (a) A feedback loop in an operational amplifier and (b) its block diagram.

The stability criterion based on the gain and phase margins can be applicable to electronic systems. Since the operational amplifier is unstable when $1 + G(s)H(s) = 0$, gain and phase margins are investigated using the condition of

$G(j\omega)H(j\omega) = -1$ or $G(j\omega) = -\frac{1}{H(j\omega)}$ in the Bode plot.

There are various types of operational amplifiers in the market, and $G(j\omega)$ is obtained from the datasheet provided by the manufacturer. One example of this data sheet of $G(j\omega)$ is shown in Figure 6.41. When $H(j\omega)$ is determined by Equation (6.33) using selected R_1 and R_f, the stability of the operational amplifier can be investigated using the condition of $G(j\omega) = -\frac{1}{H(j\omega)}$. When $R_1 \gg R_f$, the operational amplifier is unstable because of the negative gain and phase margins.

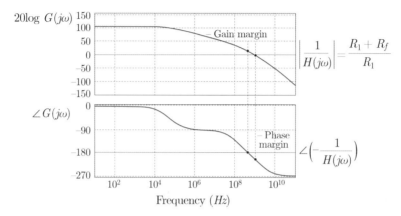

Figure 6.41: Stability investigation of an operational amplifier using $G(j\omega) = -\frac{1}{H(j\omega)}$.

6.8   Sampling effect on control

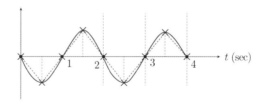

Figure 6.42: Sinusoidal signal sampled at different frequencies ($\bullet$) and ($\times$) indicate 1 and 0.5 Hz sampling frequencies, respectively.

It is usually good to have fast sampling in digital control because signal loss does not occur in this case [44]. When the 0.5 Hz sinusoidal signal shown in Figure 6.42 needs to be detected by a sensor, it must wait for at least 2 seconds to collect the data for recognizing the signal. If the signal is sampled every second, which is denoted using dot marks (●) in Figure 6.42, the sampling frequency is $1\ Hz$. At this frequency, the sinusoidal signal can be recognized as a constant value, as shown in Figure 6.42. If the signal is sampled every 0.5 s, denoted using cross marks (×) in Figure 6.42, it is recognized as a ramp signal. Therefore, it is recommended to sample a sensor at a rate faster than the sampling frequency, at least 5 ∼ 10 faster, to avoid signal distortion [45]. However, it is not always true that a high sampling frequency is always good for system control because the high-frequency-sampled signal can excite the plant and can cause vibrations due to the resonance effect. Moreover, a high sampling frequency sometimes makes it difficult to realize real-time digital control because of the large amount of data.

6.9   Control applications

6.9.1   Position control of a nano scanner

Figure 5.28 shows the configuration of a two-axis driven VCM nano scanner that was introduced in Section 5.7. The frequency response of the nano scanner for horizontal axis motion is experimentally obtained and its result is shown in Figure 6.43 [27].

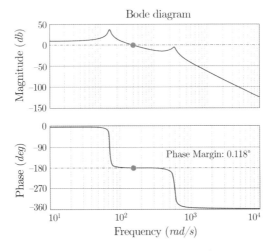

Figure 6.43: The frequency response of a VCM nano scanner.

The nano scanner transfer function, $G(s)$, is obtained using a curve fitting technique, which provides

$$G(s) = \frac{5.67 \times 10^9}{s^4 + 51.4s^3 + 3.5710^5 s^2 + 1.23910^6 s + 1.88510^9} \quad (6.34)$$

As shown in Figure 6.43, the phase margin is very small. A PD controller with $K_p(1 + T_d s)$ is used so that the phase and gain margins are properly adjusted. For example, if controller gains K_p and T_d are selected to be 0.2 and 0.01, respectively, $C(s)G(s)$ is shown in Figure 6.44. $C(s)G(s)$ has a phase margin of 47.7° and gain margin of 3.3 db. Due to the small gain margin, it is not possible to increase K_p for better performance in the closed loop system unless the second resonance peak is reduced. Hence, it is important to design a system without sub-resonance.

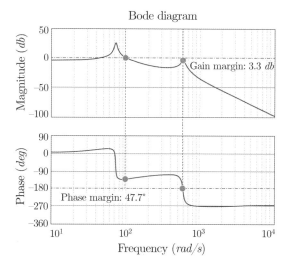

Figure 6.44: $C(s)G(s)$ with a PD controller $K_p(1 + T_d s)$, where $K_p = 0.2$ and $T_d = 0.01$

Q9: How big is the bandwidth of the closed-loop system for the $C(s)G(s)$, shown in 6.44, from the design rule-of-thumb? Check if the closed loop bandwidth is assured in the time domain using simulation.

When a PI controller with $K_p(1 + 1/(T_i s))$ is used having $K_p = 0.02$ and $T_i = 0.03$, $C(s)G(s)$ is shown in Figure 6.45. This controller produces large phase margins 93.3° and 114° respectively at $2\,rad/sec$ and $73.2\,rad/sec$. However, it has a low gain margin of 5.12 dB at $77.3\,rad/sec$. Due to the low gain crossover frequency around $2\,rad/sec$, the closed-loop bandwidth is expected to be much lower than that of a PD controller.

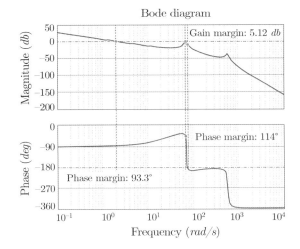

Figure 6.45: $C(s)G(s)$ with PI controller $K_p(1 + 1/(T_i s))$, where $K_p = 0.02$ and $T_i = 0.03$.

Q10: How big is the bandwidth of the closed-loop system for the $C(s)G(s)$ shown in 6.45?

6.9.2 Force control of an atomic force microscope(AFM)

The AFM is a very useful instrument for measuring the topology of nano-scale materials in various industry and research areas, such as micro-electromechanical systems (MEMS) and biology. In addition, the AFM can be used to measure physical, electrical, and magnetic properties, in addition to topological images [30].

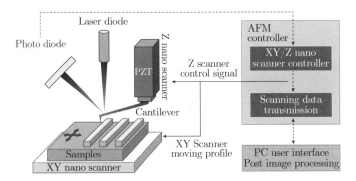

Figure 6.46: Configuration of an AFM system.

Figure 6.46 shows the configuration of the Nano Xpert, an AFM system built by EM4SYS [30]. The main components of the AFM are the laser diode, photo diode, XY and Z axes nano scanners, and cantilever. As the tiny cantilever tip approaches and makes contact with the surface of a material using the piezo-electric transducer (PZT) driven Z nano scanner, it experiences attractive and repulsive forces depending on the gap displacement between the cantilever tip and the surface, as shown in Fig 6.47. The attractive and repulsive forces deflect the cantilever tip such that they can be indirectly measured through the cantilever rotational angle. This angle is transformed into the displacement, which is conventionally measured using an optical sensor composed of a laser diode and a photo diode. Hence, the cantilever can be considered a displacement-amplification lever. The repulsive force is used for contact-type AFM measurement whereas the attractive force is used for non-contact AFM measurement. The atomic force generated due to the cantilever deflection is controlled to be constant to obtain the sample height. Furthermore, when the XY nano scanner moves the cantilever tip in the x and y directions, the three-dimensional (3D) height information of a sample can be measured over the scanned area.

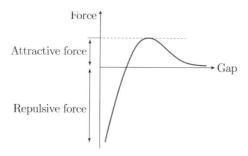

Figure 6.47: Attractive and repulsive forces depending on the gap displacement between the cantilever tip and the sample surface.

Suppose a tiny cantilever tip is in contact with a material surface whose height is z_{s1} at Location 1, as shown in Figure 6.48. The cantilever tip can be treated as a flexible spring having low stiffness. When the cantilever attached to the PZT actuator is elongated to z_1 by the supplied voltage, a repulsive force is generated in the cantilever. Then, the repulsive force bends the cantilever. If the bent angle of the cantilever is proportional to the cantilever height Δz, the repulsive force can be measured using the optical sensor mounted on the AFM head, instead of a force sensor. When the cantilever tip moves to Location 2, where the sample height is z_{s2}, the PZT actuator is elongated to z_2 using force

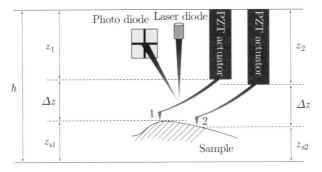

Figure 6.48: The PZT elongation depending on different sample heights.

control to have the same Δz as in Location 1. Then, sample heights z_{s1} and z_{s2} comprise the following relation:

$$h = z_{s1} + \Delta z + z_1 = z_{s2} + \Delta z + z_2 \qquad (6.35)$$

where h is the constant length between the top of the PZT actuator and the bottom of the sample. From Equation (6.35), z_{s1} and z_{s2} are respectively obtained as

$$z_{s1} = h - (\Delta z + z_1) \qquad (6.36)$$

$$z_{s2} = h - (\Delta z + z_2) \qquad (6.37)$$

Expanding the above expression to all locations on the surface, we obtain the following relation:

$$z_{si} = h - (\Delta z + z_i) \quad for \ i = 1, 2, 3, \cdots, n$$

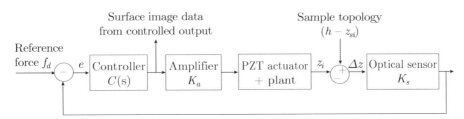

Figure 6.49: AFM control block diagram for z-axis force control.

Figure 6.49 shows a block diagram of force control used for an AFM. The desired repulsive force of the deflected cantilever, F_d is a force-control input. A voltage amplifier with a gain K_a is used for supplying a high voltage to the PZT actuator, which produces a displacement of the PZT actuator connected to one end of the cantilever. The other end of the cantilever is in contact with the sample surface. The force produced by the cantilever deflection Δz is measured by an optical displacement sensor with a gain K_s. The relative sample surface $(h - z_{si})$, which has the sample topology information, can be considered as a disturbance input.

Using the force controller, Δz is controlled to be constant. Hence, the height of the sample, z_{si}, can be obtained by measuring the elongated displacement of the PZT actuator, z_i from Equations (6.36) and (6.37). However, it is difficult to install an additional sensor to measure z_i in such a small space alongside the existing optical sensor which measures Δz. Instead, z_i can be indirectly measured by the controlled output under the assumption that there are no dynamics in the amplifier, PZT actuator, and plant, as shown in Figure 6.49. This is because z_i is proportional to the controller output at these conditions. Refer to https://www.ntmdt-si.com/resources/spm-principles/atomic-force-microscopy/ contact-afm/constant-force-mode to understand the working principle with a video demonstration.

Figure 6.50 shows a topology image of SiC graphene and its profile line $a - a'$. This result shows that the height of a few nanometers can be measured using the AFM.

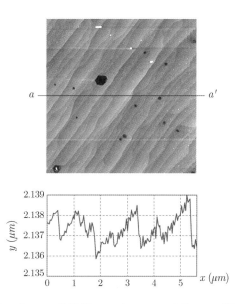

Figure 6.50: Topology of SiC graphene and its profile line $a - a'$.

6.9.3 Controller design of an unstable system

Suppose a steel ball is levitated using electromagnetic force F_{em}, as shown in Figure 6.51 [46]. The mechanical equation is obtained using Newton's law where the external forces are the same as the inertia force.

$$F_{em} - mg = -m\ddot{x} \tag{6.38}$$

Here, F_{em} is a function of applied current i and distance x. F_{em} is experimentally obtained as shown in Figure 6.52.

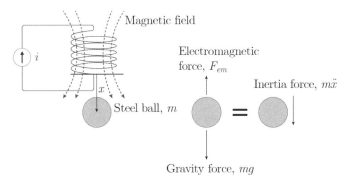

Figure 6.51: A steel ball levitated using magnetic force.

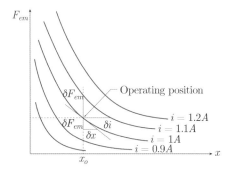

Figure 6.52: Electromagnetic force variation with respect to current and position.

Under the assumption that the motion is very small, F_{em} can then be linearly represented at the equilibrium position and current, x_o and i_o, respectively, as

$$F_{em} = K_1 x + K_2 i \tag{6.39}$$

K_1 and K_2 can be respectively obtained by

$$K_1 = \frac{\delta F_{em}}{\delta x}\Big|_{i=i_o} \tag{6.40}$$

$$K_2 = \frac{\delta F_{em}}{\delta i}\Big|_{x=x_o} \tag{6.41}$$

Note that K_1 is negative. Subsequently, using Equation (6.38) and (6.39), we obtain

$$m\ddot{x} + K_1 x + K_2 i = mg \tag{6.42}$$

where i can be considered the summation of constant current i_o (for holding the steel ball in the equilibrium position) and applied current $-i'$ (for controlling the ball). x is similarly considered as the summation of x_o and x'. Thus, Equation (6.42) is modified as

$$m\ddot{x}' + K_1(x_o + x') + K_2(i_o - i') = mg \tag{6.43}$$

When $i_o = \frac{(mg - K_1 x_o)}{K_2}$, the transfer function of the levitating ball, $G(s)$ is represented by the input i' and output x' to obtain

$$G(s) = \frac{x'(s)}{i'(s)} = \frac{K_2}{(ms^2 + K_1)} \tag{6.44}$$

The transfer function represented in Equation (6.44) has poles in the right half plane because K_1 is negative. Hence, the levitating ball is unstable. Figure 6.53 shows a control block diagram of the levitating ball which makes the plant stable. Here, V_d and V_x are the command input voltage and output position voltage, respectively.

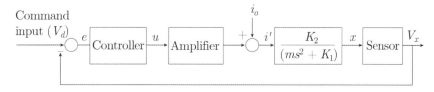

Figure 6.53: Control block diagram of the levitated ball.

Q11: What types of controllers can be used to make the levitated ball stable in the feedback system? Prove a stable feedback system based

on the pole location of the closed loop system.

6.10 Loop shaping controller

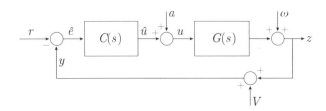

Figure 6.54: A closed loop system with disturbances and noise.

Differently from the ideal closed loop system in which disturbances and noise are not considered as shown in Figure 6.17, there exists several input disturbances and sensor noise in a real closed loop system, as shown in Figure 6.54. Here, a and w are the input and output disturbances, respectively. The frictional force and external vibration are examples of a and w, respectively, in a mechanical system. Electric noise is one example of v. r is the command input, z is the output signal, and $\hat{e}$ is the error, which is defined as the difference between r and contaminated output y. $\hat{u}$ is the controlled output and u is the plant input or contaminated controlled output. Finally, we obtain the following relations:

$$\hat{e} = r - y = r - (z + v) \tag{6.45}$$

$$u = \hat{u} + a = C\hat{e} + a \tag{6.46}$$

$$z = Gu + w \tag{6.47}$$

$$y = z + v \tag{6.48}$$

u is rewritten by using Equation (6.45) as

$$u = C(r - z - v) + a \tag{6.49}$$

Hence, z is obtained using Equation (6.47) as

$$z = CG(r - z - v) + Ga + w \tag{6.50}$$

Rearranging Equation (6.50), we obtain

$$z = (I + CG)^{-1} CG (r - v) + (I + CG)^{-1} Ga + (I + CG)^{-1} w \qquad (6.51)$$

Then, the true error signal, e, which is defined as $(r - z)$, is given by

$$
\begin{aligned}
e &= r - z \\
&= \{I - (I + CG)^{-1}CG\} r + (I + CG)^{-1}\{CGv - Ga - w\}
\end{aligned}
$$

Considering that $I - (I + CG)^{-1}CG = (I + CG)^{-1}$ and $(I + CG)^{-1}CG = CG(I + CG)^{-1}$, we obtain

$$e = (I + CG)^{-1}r + CG(I + CG)^{-1}v - (I + CG)^{-1}Ga - (I + CG)^{-1}w \quad (6.52)$$

where $(I + CG)^{-1}$ is a sensitivity transfer function S and $CG(I + CG)^{-1}$ is a complementary transfer function T [47]. Then, we obtain

$$S + T = 1 \qquad (6.53)$$

e is then simply represented using the definitions of S and T as

$$e = S(r - Ga - w) + Tv \qquad (6.54)$$

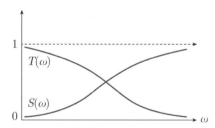

Figure 6.55: Frequency-dependent relation between $S(\omega)$ and $T(\omega)$ to reduce e.

There are four variables r, a, w and v in Equation (6.54). One thing to note here is that the four variables have frequency-dependent properties. In general, r, a, and w are much larger than v at a low frequency while v is much larger than r, a, and w at a high frequency. As the goal of feedback control is to minimize e as quickly as possible, $S(\omega)$ should be maintained to be low at a low frequency to reduce e. Though $T(\omega)$ is high at a low frequency, a high $T(\omega)$

does not contribute to increasing e because v is very small at a low frequency. Similarly, $S(\omega)$ should be maintained to be high at a high frequency to make $T(\omega)$ small, which contributes to decreasing the effect of v that is dominant at a high frequency. Therefore, S and T should have the frequency-dependent magnitude variation represented in Figure 6.55.

Controller implementation using operational amplifiers

—

Controller implementation using operational amplifiers

We now implement a real feedback controller to construct the closed loop system configured in Figure 6.36. A conventional controller, such as a PID and a compensator, can be easily implemented using either analog electronic circuits or digital processors. Herein, analog electronic circuits are implemented to be more consistent with an analog plant (system) and for easy control simulation. In addition, voltage and current amplifiers are studied for realistic controller implementation in a mechanical system.

7.1   Load-effect-free operational amplifier

When electronic circuits composed of passive components are used for constructing an analog controller, one thing to note is that the part following electronic circuits can cause a load effect on the previous part. This happens because a circuit composed of passive components is a non-isolated system, as addressed in Section 1.6.

An operational amplifier can be used to take advantage of the no-load effect via electronic feedback control. This amplifier has a high input impedance condition; particularly, it has a very large R_∞. Hence, current i_2 is prevented from flowing into the next circuit, which is represented with a dotted line in Figure 7.1. This is why a front circuit comprising R_1 and C_1 can be considered to be isolated from the next circuit comprising R_2 and C_2. Additionally, the amplifier has a very low output impedance condition; particularly, it has a very low R_s. Hence, the current from the power supply can flow well into the next circuit. The operational amplifier circuit used for this purpose is called a buffer.

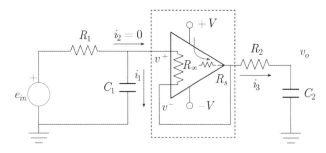

Figure 7.1: A high input impedance and low output impedance device.

It is possible to develop several isolating systems using operational amplifiers that contain unit gains (or buffers) between the passive systems, as shown in Figure 7.2. Hence, there is no need to do circuit analysis again when a new passive circuit is added to the initial circuit. This is a remarkable advantage of using operational amplifiers.

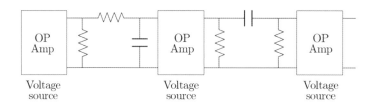

Figure 7.2: Operational amplifiers between passive systems to make them be isolated.

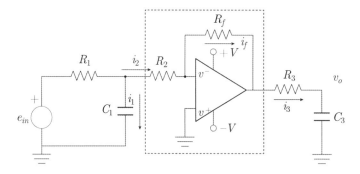

Figure 7.3: An inverting amplifier used for isolating the electronic circuits.

Q1: The inverting amplifier shown in Figure 7.3 is an example of a

real operational amplifier having a gain of $\frac{R_f}{R_2}$ which isolates the front circuit composed of R_1 and C_1 and the next passive circuit composed of R_3 and C_3. What is the condition of R_2 in order to achieve this purpose? (Hint: i_2 should be as small as possible to isolate the front circuit.)

7.2 Amplifiers

7.2.1 Voltage amplifier

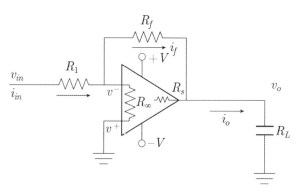

Figure 7.4: An operational voltage amplifier.

An ideal voltage source can be made using an operational amplifier that has high input impedance (resistance) R_∞ and low output impedance (resistance) R_s. Figure 7.4 shows a schematic diagram of an operational voltage amplifier wherein the output voltage, v_o, is generated by the input voltage, v_{in}, with a proportional gain regardless of the amount of current flowing through load resistor R_L. This amplifier has v^- and v^+ inputs connected to v_{in} and ground, respectively. A power supply is connected to the $+V$ and $-V$ inputs. The positive voltage, $+V$, and the negative voltage, $-V$, are the maximum and minimum voltages, respectively, that can be supplied to the operational voltage amplifier. Hence, v_o cannot exceed these voltage values.

Due to the condition of the high input impedance of the operational amplifier, namely $R_\infty \cong \infty$, input current i_{in} cannot flow through the resistor but must flow through feedback resistor R_f. Thus, the feedback loop makes output current i_f equal to i_{in}. These conditions lead to the following results:

$$v^+ = v^- = 0 \tag{7.1}$$

$$i_{in} = \frac{v_{in} - v^-}{R_1} = \frac{v_{in}}{R_1} \tag{7.2}$$

$$i_f = \frac{v^- - v_o}{R_f} = -\frac{v_o}{R_f} \tag{7.3}$$

$$i_{in} = i_f \tag{7.4}$$

From the above equations, we obtain

$$v_o = -\frac{R_f}{R_1} v_{in} \tag{7.5}$$

v_o is generated by v_{in} with the following proportional gain, $-\frac{R_f}{R_1}$. Then, the current i_o drawn from the power supply flows through R_L. i_o is determined only by R_L because it has a low output impedance, namely $R_s \cong 0$.

Q2: What is the voltage across $R_L = 1\ k\Omega$ when the voltage source 10 V is applied to the circuit shown Figure 3.6 for $R_{in} = 50\ \Omega$? How is the the voltage across $R_L = 1\ k\Omega$ changed when $R_{in} = 1\ k\Omega$?

Q3: Figure 7.5 shows a non-inverted amplifier. Use the circuit analysis to obtain the input and output relation.

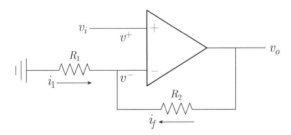

Figure 7.5: A non-inverted amplifier.

A voltage amplifier can be used for supplying high voltages to electrostatic or piezoelectric actuators for rotational and linear scanning actuators, respectively [48].

7.2.2 Current amplifier

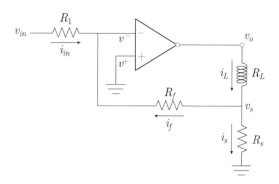

Figure 7.6: An operational current amplifier.

Figure 7.6 shows a current source used for amplifying current proportionally generated by the input voltage regardless of the magnitude of load resistor, R_L [49]. There is a feedback loop through the feedback resistor, R_f. Due to the high input impedance condition of the operational amplifier, input current, i_{in} cannot flow through its resistor but instead flows through the feedback resistor, R_f. Thus, the feedback loop makes output current i_f equal to the input current i_{in}. The load current, i_L, is divided into feedback current i_f and i_s. The above conditions lead to the following results:

$$v^+ = v^- = 0 \tag{7.6}$$

$$i_{in} = \frac{v_i - v^-}{R_1} = \frac{v_{in}}{R_1} \tag{7.7}$$

$$i_f = \frac{v_s}{R_f} = -i_{in} \tag{7.8}$$

$$i_L = i_f + i_s = \frac{v_s}{R_f} + \frac{v_s}{R_s} = v_s \left(\frac{1}{R_f} + \frac{1}{R_s} \right) \tag{7.9}$$

$$v_s = R_f i_f = -R_f i_{in} = -R_f \left(\frac{v_{in}}{R_1} \right) = -\frac{R_f}{R_1} v_{in} \tag{7.10}$$

From the above equations, we obtain the relation that output current, i_L is amplified by input voltage, v_{in} as

$$i_L = -\frac{R_f}{R_1} \left(\frac{1}{R_f} + \frac{1}{R_s} \right) v_{in} = -\frac{R_f}{R_1} \frac{(R_s + R_f)}{R_f R_s} v_{in} = -\frac{R_s + R_f}{R_1 R_s} v_i \tag{7.11}$$

R_s is used to flow the current to the ground direction because R_f is usually chosen to be much larger than R_L. Hence, $i_s \cong i_L$. If R_s does not exist, the high current output will flow to R_f, which will burn out the operational amplifier. R_s is also used to monitor i_L in terms of voltage. The power consumption in R_s is calculated as $i_s^2 R_s$. Therefore, the material property of R_s should be strong enough that it is not burned.

A current amplifier can be used for supplying high current to electromechanical actuators, such as electrical motors or voice coil motors(VCMs), respectively for rotational or linear actuation [50],[51].

7.3 Various controllers using operational amplifiers

Table 7.1: Various filters constructed using operational amplifiers.

	Controller	$G(s) = \dfrac{E_o(s)}{E_i(s)}$	Operational amplifier circuit
1	P	$-\dfrac{R_2(s)}{R_1(s)}$	
2	I	$\dfrac{R_4}{R_3}\dfrac{1}{R_1 C_2 s}$	
3	PD	$\dfrac{R_4}{R_3}\dfrac{R_2}{R_1}(R_1 C_1 s + 1)$	
4	PI	$\dfrac{R_4}{R_3}\dfrac{R_2}{R_1}\dfrac{(R_2 C_2 s+1)}{R_2 C_2 s}$	
5	PID	$\dfrac{R_4}{R_3}\dfrac{R_2}{R_1}\dfrac{(R_1 C_1 s+1)(R_2 C_2 s+1)}{R_2 C_2 s}$	
6	Lead or Lag	$\dfrac{R_4}{R_3}\dfrac{R_2}{R_1}\dfrac{(R_2 C_2 s+1)}{(R_1 C_1 s+1)}$	
7	Lag-lead	$\dfrac{R_6}{R_5}\dfrac{R_4}{R_3}\dfrac{[(R_1+R_3)C_1 s+1](R_2 C_2 s+1)}{(R_1 C_1 s+1)[(R_2+R_4)C_2 s+1]}$	

Analog electronic circuits with operational amplifiers have been used to construct feedback controllers, some of which are listed in Table 7.1 [37]. These circuits are all analog linear controllers. Of course, they can be implemented digitally in a computer or microprocessor for more efficient controller use and to save the time required to choose the right electronic components for resistors and capacitors. However, a feedback controller using analog electronic circuits help us understand physical systems and their behavior more effectively and quickly.

To implement an analog feedback controller, the difference between the command input and output must be fed into the controller, as shown in Figure 6.36, so a subtractor circuit shown in Figure 7.7(a) is needed for a negative feedback loop. Since the signal is inverted each time it passes through an inverting operational amplifier, an adder circuit shown in Figure 7.7(b) can also be used if the output signal is inverted. Therefore, we need to be careful when selecting subtractor and adder circuits for these feedback loops. Otherwise, a positive feedback loop can be implemented, which results in the instability of the control system.

The relations between the inputs e_{i1}, e_{i2} and output e_o are represented respectively in Equations (7.12) and (7.13) for the subtractor and adder, respectively:

$$e_o = \frac{R_3}{R_2 + R_3}(1 + \frac{R_f}{R_1})e_{i1} - \frac{R_f}{R_1}e_{i2}, \tag{7.12}$$

$$e_o = -(\frac{R_f}{R_1}e_{i1} + \frac{R_f}{R_1}e_{i2}) \tag{7.13}$$

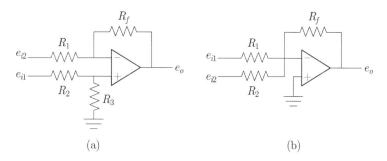

(a)   (b)

Figure 7.7: (a) Electronic subtractor and (b) adder using operational amplifiers.

Figure 7.8 shows the configuration of position feedback control of the VCM nano scanner shown in Figure 5.28, including the experimental instruments required for its feedback control. A function generator is used for applying a ramp input signal to the VCM scanner for forward and backward scanning in

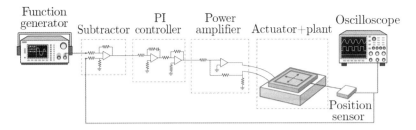

Figure 7.8: Configuration of feedback control of the VCM nano scanner and experimental instruments required for its feedback control.

the lateral directions. An analog PI controller is used for improving the position precision. The current amplifier is constructed for applying $0 \sim 1\,A$ current to the electromagnetic actuator. A position sensor is required to feedback on the VCM scanner position signal to the input. A subtractor is constructed using the analog circuit shown in Figure 7.7, and an oscilloscope can be used to measure output signals for obtaining the control performance.

7.4 Electronic circuits for a plant realization and feedback control

A real mechanical plant can be replaced using an equivalent electronic circuit composed of an integrator, low-pass filters, and high-pass filters to simulate control performance in advance. They are introduced below.

7.4.1 Integrator

Figure 7.9: Integrator in a mechanical system.

A mechanical system composed of mass m is shown in Figure 7.9. When the input and outputs are force f and velocity v, respectively, open loop transfer function, $G(s)$ is obtained from the dynamic equation of the mechanical system.

$$G(s) = \frac{V(s)}{F(s)} = \frac{1}{ms} \tag{7.14}$$

$G(s)$ has a form of an integrator whose frequency response is shown in Figure 4.13.

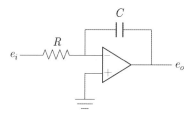

Figure 7.10: An integrator using an operational amplifier.

Figure 7.10 shows the electronic circuit of an integrator made of an operational amplifier whose transfer function is $\frac{1}{RCs}$. The input and output relation is obtained by deriving the electronic equations from circuit analysis. As a more convenient method for determining the transfer function in operational amplifiers, the impedance method can be used [52]. In this method, the transfer function is obtained by the ratio of the impedance of the feedback loop, Z_f to that of the feed-forward loop Z_i, i.e.,

$$G(s) = -\frac{Z_f}{Z_i} = -\frac{1}{RCs} \tag{7.15}$$

Using Equation (7.15), R and C can be chosen appropriately to implement m.

7.4.2 Low pass filter

A first-order mechanical system composed of mass m and damper c is shown in Figure 6.12. When the input and output are force f and velocity v, respectively, open loop transfer function $G_m(s)$ is obtained from the dynamic equation of the mechanical system.

$$G_m(s) = \frac{V(s)}{F(s)} = \frac{1}{ms + c} \tag{7.16}$$

$G_m(s)$ has a similar form as the low pass filter constructed using the electrical circuit shown in Figure 3.2.

Figure 7.11 shows the electronic circuit of a low-pass filter made with an operational amplifier whose transfer function is obtained using the impedance method.

$$G_e(s) = \frac{E_o(s)}{E_i(s)} = -\frac{\frac{1}{1/R_2 + (Cs)}}{R_1} = -\frac{R_2}{R_1(R_2Cs + 1)} \tag{7.17}$$

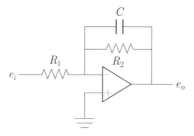

Figure 7.11: A low-pass filter using an operational amplifier.

As Equation (7.16) has a similar form to Equation (7.17), the first-order mechanical system can be realized using the low-pass filter. Moreover, when an integrator circuit is serially connected to the low-pass filter circuit, we can construct a second-order mechanical system in which the input and output are force f and displacement x, respectively, as represented by Equation (7.18).

$$G_m(s) = \frac{X(s)}{F(s)} = \frac{1}{s(ms + c)} \tag{7.18}$$

Q4: Check if Equation (7.17) is also obtained using the circuit analysis.

7.4.3 High pass filter

One electronic circuit of a high pass filter can be constructed using the circuit as shown in Figure 7.12 [52]. The transfer function of this high-pass filter is

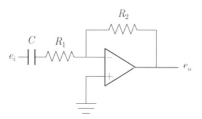

Figure 7.12: A high pass filter using an operational amplifier.

obtained as follows.

$$G(s) = -\frac{R_2 C s}{1 + R_1 C s} \tag{7.19}$$

Its frequency response is shown in Figure 4.15. To have a unity gain ratio at a high frequency, $R_1 = R_2$. The high-pass filter can be used to eliminate low-frequency signals, such as the electrical offset or DC (constant magnitude) disturbance, which mostly occur in a low-frequency range because it features a large reduction in the magnitude ratio before the cutoff frequency. Moreover, a second-order high-pass filter can be designed to further eliminate low-frequency signals due to the higher reduction rate, $-40\,db/dec$.

Q5: Check if Equation (7.19) is also obtained using the circuit analysis.

7.5 Experiment of feedback control

Suppose a mechanical system is a first-order system composed of a mass and damper as shown in Figure 6.12. Its transfer function was represented by Equation (7.16). The mechanical plant was realized using an electronic low-pass filter, as shown in Figure 7.11, whose transfer function was represented in Equation (7.17).

Figure 7.13 shows an example of an electronic circuit composed of an adder, PI controller, and electronic plant. A voltage amplifier is not implemented because a high-voltage supply is not needed. A sensor is also unnecessary because the output signal, V_{out}, can be directly measured. An adder is used instead of a subtractor to obtain the error signal, $V_{in} - V_{out}$, because V_{out} is a sign inverted signal in this case.

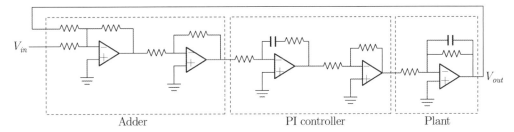

Figure 7.13: Electronic circuit diagram composed of a plant, PI controller, and adder.

Operational amplifiers are produced on electronic chips, making them convenient for industrial use. Figure 7.14 shows an electronic chip configuration of one of the inverting operational amplifiers, AD 8021 manufactured by Analog Device [54]. There are eight pins that are connected to a positive supplied voltage $+V$, a negative supplied voltage $-V$, input v^+, input v^-, output V_{out}, etc. We can construct an analog circuit following the procedures introduced by the chip datasheet. It is important to note that operational amplifiers come in a variety of chip configurations in terms of specifications and performance. Hence, we need to choose an appropriate chip for the purpose of circuit applications.

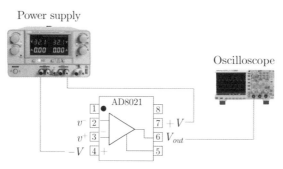

Figure 7.14: The pin configuration of one of the operational amplifiers manufactured by Analog Device.

Figure 7.15 shows a real electronic circuit implemented on a breadboard using the S/W called Fritzing [55]. $+V$, $-V$, and Ground are the points connected to the positive voltage ($+5 \sim +24$ V), negative voltage supply ($-5 \sim -24$ V), and ground. Before constructing a feedback system, it is necessary to check whether the adder, controller, and electronic plant circuits are individually working well for their intended uses. Time or frequency response analysis can be used for this purpose. However, if the plant model has an integrator, such as $\frac{1}{s(ms+c)}$, it would not be easy to check whether the corresponding circuit is well implemented because a signal easily saturates in the integrator circuit. If all components are working without problems in the open loop, the feedback loop can be connected to a command input using a subtractor or adder.

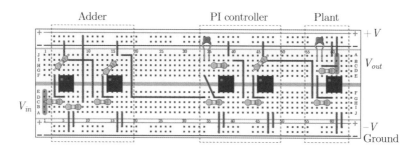

Figure 7.15: Real circuit implemented on a breadboard.

Q6: Consider the same mass–damper plant whose transfer function is $\frac{1}{ms+c}$. The mass is 0.01 kg and the damping coefficient is 2 $Nsec/m$. Components R_1, R_2, and $C_2 = 0.01\ \mu F$ are properly selected to provide a plant transfer function of $\frac{1}{0.01s+2}$. Design PI and PD controllers represented by $K_p(1 + \frac{1}{T_i s})$ and $K_p(1 + T_d s)$, respectively for velocity control referring to Table 7.1. How do the control gains affect the closed system performances in terms of speed? Compare command input and velocity output signals when the command inputs are unit step and ramp inputs.

Signal processing for
sensing systems

—

Signal processing for sensing systems

8.1   Sensor specifications

Engineers often need to design a sensor suitable for application purposes. The important performance specifications that need to be considered during sensor design are as follows: speed, precision, and linearity. It was introduced in Section 6.4.3 that speed performance can be understood as the bandwidth that can be investigated by using its frequency response. The higher the bandwidth of the sensor, the better its performance in high-speed systems because it can provide accurate signals without magnitude reduction and phase delay.

To investigate the sensor precision performance, we investigate a typical optical sensor that is composed of a laser diode (LD), lens, and photo diode (PD), as shown in 8.1. The sensor measures the height of an object using the triangulation principle. The height information, h, is measured using the signal processing of the photo-current generated in the PD. This information is finally converted to voltage because it is easier to manipulate. The laser reflecting from the surface of an object at a nominal distance, d, is focused on the center of the PD and measured by zero voltage. The maximum and minimum distances from the nominal distance, $+h$ and $-h$, are measured using the maximum and minimum voltage (for example, $10V$ and $10\ V$), respectively. When h is increased or decreased, the voltage exhibits a proportional increase or decrease due to the triangle similarity of two triangles aob and $a'ob'$

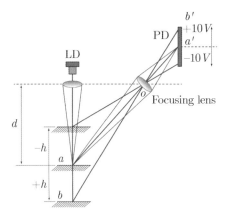

Figure 8.1: An optical sensor that uses the triangulation principle.

A one-dimensional (1D) optical sensor based on the triangulation principle can be expanded to a two-dimensional (2D) optical sensor that measures the width and height of a block. When the laser/detector pair is replaced by a line laser/2D array detector, the width and height of a block are obtained by the pixel information obtained from the 2D array detector or image sensor, as shown in Figure 8.2. This is almost the same principle that we see objects through the lens and retina of our eyes. The width and height of the block are mapped to the 2D array detector through the lens with a scale ratio.

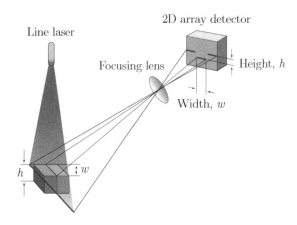

Figure 8.2: Shape of a block measured using a 2D array detector.

Suppose that Figure 8.3 shows the experimental results of the 1D sensor at the nominal distance which produces zero voltage output signal. The signal is usually contaminated with the electronic noise V_{min} coming from the power supplies and electronic devices which determines the minimum height distance,

h_{min}. V_{min} is the minimum voltage we can guarantee to call the lowest signal which is not varied for a long time regardless of the operating conditions such as temperature, humidity, and light conditions.

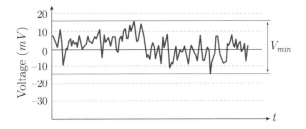

Figure 8.3: Noise level corresponding to sensor resolution.

When the maximum height distance h_{max} is measured by the maximum voltage V_{max}, we obtain the following relation:

$$h_{max} : V_{max} = h_{min} : V_{min} \tag{8.1}$$

Then, h_{min} is expressed as

$$h_{min} = \frac{V_{min}}{V_{max}} h_{max} \tag{8.2}$$

For example, if $V_{max} = 10$ V, $V_{min} = 10$ mV, $h_{max} = 10$ cm, then $h_{min} = 0.1$ mm. This result indicates that the sensor can measure a height as small as 0.1 mm, which is called sensor resolution. If $hmax$ is adjusted to measure a smaller length of 1 μm, then the resolution is improved to measure up to $h_{min} = 1$ nm. Can you imagine that such a small length can be measured using an ordinary optical sensor? According to Equation (8.2), it looks like it would be preferable to increase V_{max} to improve the resolution. However, increasing V_{max} causes an increase of V_{min}, which yields the same resolution. Therefore, to lower the resolution, the signal-to-noise ratio (S/N) should be kept large. A sensor with a small resolution is called a high-precision sensor. Several signal processing techniques will be introduced in Sections 8.3 and 8.4 to increase the S/N ratio.

The other sensor performance parameter is linearity. Figure 8.4 shows experimental results for how the sensor output represents height distance. If a sensor having good linearity has the relation $V = K_s h$, the height is correctly obtained by calculating the sensor output voltage divided by constant K_s. However, if a sensor has a non-linearity characteristic as represented by the dotted line, the height can be compensated to h_{real} by considering h_{comp} which differs from the voltage signals as

$$h_{real} = h - h_{comp}(V) \tag{8.3}$$

The nonlinearity of a sensor can be compensated mathematically using a post-processing technique. One of these compensation methods is called the lookup table method [56].

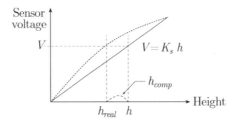

Figure 8.4: Sensor linearity.

8.2 Components of optical sensors

8.2.1 Laser diode of a light source

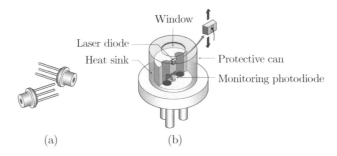

Figure 8.5: (a) Photograph and (b) internal configuration of a semiconductor LD.

For a light-source component for an optical sensor, the use of LDs has dramatically increased recently owing to their small size, low power consumption, and ruggedness [57]. Coherence and single-wavelength characteristics of LDs enable the laser beam to be focused to a diffraction-limited spot size. Another advantage of LDs is that the laser beam can be directly modulated at high frequencies, up to several gigahertz for high-speed data communications. The external and internal configurations of a standard LD are respectively shown in Figures 8.5(a) and 8.5(b).

To drive an LD for light generation, it is necessary to supply it with a certain amount of current. Figure 8.6(a) shows an electronic circuit of a current amplifier for an LD that uses a transistor [58]. Here, the LD is modeled as R_c. There are three legs, called a base (B), emitter (E), and collector (C), in a transistor whose real configuration is shown in Figure 8.6(b). Here, V_{BB} is the supply voltage to the base, V_{CC} is the supply voltage to the collector, and R_b is the base resistor

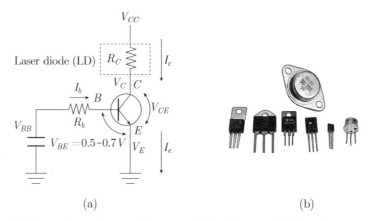

(a)   (b)

Figure 8.6: (a) Transistor for current amplification and (b) real transistor configuration.

The transistor controls the amount of current I_c flowing through R_c by the amount of current I_b flowing through R_b. Thus, we obtain

$$I_c = \beta I_b \tag{8.4}$$

Here, β is the current amplifying gain, which is $200 \sim 300$. Additional electronic circuit equations associated with Figure 8.6 are derived as

$$V_{BB} = R_b I_b + V_{BE} \tag{8.5}$$
$$I_e = I_c + I_b = (\beta + 1)I_b \tag{8.6}$$
$$V_{CC} = R_c I_c + V_{CE} \tag{8.7}$$

Here, V_{BE} is the voltage drop for its operation between the base and emitter, which is $0.5 \sim 0.7V$. V_{CE} is the voltage drop between the collector and emitter. From Equation (8.5), we obtain

$$I_b = \frac{V_{BB} - V_{BE}}{R_b} \tag{8.8}$$

When I_b is determined from Equation (8.8), the amplified current I_c flows in proportion to β through the laser diode. The relation of $I_b - I_c$ with respect to V_{CE} is shown in Figure 8.7. I_c increases proportionally to V_{CE} at a voltage less than $0.7\ V$ and is saturated to a constant current at a voltage larger than $0.7\ V$. As shown in Figure 8.7, there is no big constraint on V_{CE} for a good operation of the transistor as the almost constant current can be obtained regardless of its magnitude when a proper range of R_c is selected.

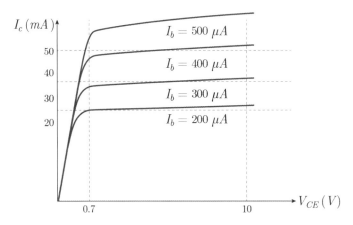

Figure 8.7: Characteristic curve of a transistor.

For example, if you want to run $I_c = 100\ mA$ on an LD, we can determine how large of resistance R_b is needed. First, assuming the current gain β is around 100, I_b that should flow to the base is $1\ mA$ from Equation (8.4). When V_{BB} is 5 V, and V_{BE} is 0.7 V, the voltage across R_b is 4.3 V. Hence, $R_b = 4.3\ V/0.001\ A = 4.3\ k\Omega$.

We can also find out a mechanical device that is equivalent to the transistor. Figure 8.8 shows servo valves used in a hydraulic system. If the spool moves slightly away from the neutral position using a small electromagnetic force, the flow increases greatly in proportion to the opened area of the spool valve and enters the cylinder to move the piston. Just as a very low current in the base can generate a large current in the collector, a very small movement in the spool

x_v can generate a large flow, Q_1.

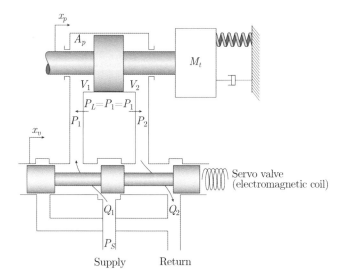

Figure 8.8: A servo valve in a hydraulic system that is equivalent to an electronic transistor.

Q1: When $10~V$ V_{CC} is used, select an appropriate resistance R_c to make the transistor shown in Figure 8.6 work well for current amplification. Measure V_{CE} using an oscilloscope.

8.2.2 Photo diode for a detector

The light or laser emitted from the LD is reflected from the surface of an object and it is received by a PD. A PD is a semiconductor device that converts light into electrical current through the photoelectric effect [59].

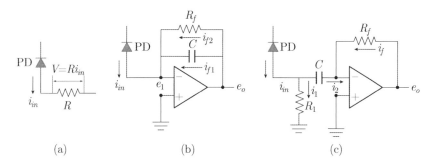

Figure 8.9: Several TIA circuits, having (a) only amplification, (b) amplification with a low pass filter, and (c) amplification with a high pass filter.

When current i_{in} is generated from the PD, the current can be simply converted to voltage V using a passive circuit, in which the PD is connected to resistor R as shown in Figure 8.9(a). Then, the voltage is obtained by the relation $V = Ri_{in}$. Figures 8.9(b) and 8.9(c) show active electronic circuits that convert the photo current to a voltage output, which are called trans-impedance amplifiers (TIA) and are almost exclusively implemented with one or more operational amplifiers. High and low-pass filters are accordingly added for noise and DC (constant) offset rejection as well as for current–voltage amplification.

The electronic equations of the TIA shown in Figure 8.9(b) are derived using the characteristics of the operational amplifier as

$$I_{in} + I_{f1} + I_{f2} = 0 \tag{8.9}$$

$$I_{f1} = (E_o - E_1)Cs \tag{8.10}$$

$$I_{f2} = \frac{E_o - E_1}{R_f} \tag{8.11}$$

$$E_1 = 0 \tag{8.12}$$

where all variables and parameters are graphically defined. Then, the transfer function of the current-voltage relation is obtained using Equations (8.9) through (8.12).

$$\frac{E_o(s)}{I_{in}(s)} = -\frac{R_f}{R_f Cs + 1} \tag{8.13}$$

As expressed in Equation (8.13), output voltage E_o is amplified by a factor of R_f at a low frequency. Moreover, the high-frequency noise of E_o is filtered out because Equation (8.13) is a low pass filter. Of course, a TIA with a high pass filter can be constructed as shown in Figure 8.9 (c) to remove low-frequency disturbances, such as a DC (constant) offset signal.

Q2: Derive the transfer function of the TIA circuit shown in Figure 8.9(c).

8.3 Amplitude modulation

When a constant intensity laser beam from a laser diode (LD) falls on a photo diode (PD), it produces a constant direct current (DC) proportional to the laser power or location. However, the DC may contain dark current [60], which flows through the photo detector even when no laser beam enters the device. Additionally, other background light sources, such as sunlight or fluorescent light, can generate unwanted current in a PD when it is exposed to such lights. These types of currents deteriorate sensor accuracy and precision.

Modulation and demodulation techniques [61] can be used for improving sensor precision and they have been applied to industrial areas, such as information, communication, and electronics. Some typical signal processing techniques are introduced here with an emphasis on their real applications to sensing systems.

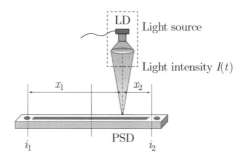

Figure 8.10: An optical sensor composed of an LD and PSD.

Figure 8.10 shows an optical sensor composed of an LD and a position-sensitive detector (PSD), which is one type of PDs. When the laser beam falls on the PSD, the laser beam position, x_1 and x_2 are measured using currents i_1 and i_2 generated at both ends of the PSD [60] having relations of

$$i_1 \propto \frac{1}{x_1}, \quad i_2 \propto \frac{1}{x_2}$$

Additionally, there exist external light disturbances, such as sunlight or fluorescent light, that can cause measurement errors and electronic noise. We can solve these problems using modulating techniques.

One technique is the amplitude modulation technique, which is frequently applied to an LD. When an LD is sinusoidally modulated by current having frequency ω_c, the intensity, $I(t)$, of the laser beam can be expressed as

$$I(t) = I_o + m \, \sin(\omega_c \, t) \, (I_o > m) \tag{8.14}$$

Here, DC (constant) offset, I_o is mathematically added in Equation 8.14 to prevent the modulated intensity from being negative. m is the amplitude of the sinusoidal modulation. When the intensity of the external disturbances, such as ambient light, is $\zeta(t)$, the current i_1 obtained from the PSD is represented as [60],[62]

$$
\begin{aligned}
i_1 &= K[I(t) + \zeta(t)]\frac{1}{x_1} \\
&= K[I_o + m\ \sin(\omega_c\ t + \alpha) + \zeta(t)]X_1 \quad\quad (8.15)
\end{aligned}
$$

Here, K is a proportional constant, and α is the phase lag due to the PSD material characteristics and processing electronic circuit. X_1 is defined as $\frac{1}{x_1}$ to express the relation in a simple form.

Considering that I_o is constant and $\zeta(t)$ is the low-frequency disturbance, and ω_c is a high frequency, we can remove I_o and $\zeta(t)$ in the signal, $i(t)$, using a high pass filter or band pass filter centered at ω_c. Then, the resulting signal of the PSD sensor contains only the current induced by the sinusoidally driven LD light source, i.e.,

$$
i_1 = Km \cdot X_1 \sin(\omega_c\ t + \alpha) \quad\quad (8.16)
$$

Comparing Equations (8.15) and (8.16), it is observed that the unwanted DC (constant) values and the low-frequency disturbances are eliminated and only the position information of X_1 remains. This advantage is obtained because the position signal X_1 is modulated by the sinusoidal signal with the carrier frequency ω_c so that a band-pass filter or a high-pass filter can be applied. When a higher frequency, ω_c, is chosen, the satisfactory decreasing effect of low-frequency disturbances can be obtained.

Equation (8.16) is the standard form of amplitude modulation, which places the term X_1 as the amplitude of $\sin(\omega_c t + \alpha)$. Hence, it is called amplitude modulation. X_1 contains information that we want to know. Figure 8.11 shows the modulated signal i_1 when X_1 is in slow motion and represented by $X_1 = \sin(2\pi t)$, $\omega_c = 2\pi \cdot 50Hz$, $\alpha = 0$ and $Km = 1$.

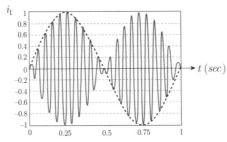

Figure 8.11: Graph of i_1 when $X_1 = \sin(2\pi t)$, $\omega_c = 2\pi \cdot 50Hz$, and $Km = 1$.

To understand the amplitude modulation process in the frequency domain, we can simply rewrite $i(t)$ with the assumptions $\alpha = 0$ and $X(t) = \sin \omega_m t$ as follows:

$$i_1 = Km \sin(\omega_m \ t) \cdot \sin(\omega_c \ t) = -\frac{1}{2} Km[\cos(\omega_c + \omega_m)t - \cos(\omega_c - \omega_m)t] \quad (8.17)$$

i_1 in the frequency domain indicates that the frequency of X_1 is shifted to $\omega_c + \omega_m$ and $\omega_c - \omega_m$, as shown in Figure 8.12. In other words, the frequency of the position signal is shifted to both bands of the driving frequency, ω_c. In conclusion, the amplitude modulation technique provides the advantage of shifting the frequency of the motion of X_1 to $\omega_c - \omega_m$ and $\omega_c + \omega_m$, where external disturbances of X_1 are at a minimum and ω_c is used to separate the low-frequency external disturbances of X_1. As a result, unwanted DC (constant) values and low-frequency disturbances can be easily eliminated using high-pass filters or band-pass filters.

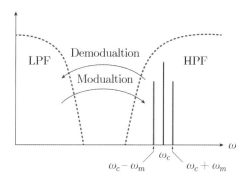

Figure 8.12: Frequency of X_1 shifted to $\omega_c + \omega_m$ and $\omega_c - \omega_m$ where external disturbances of X_1 are minimum.

Q3: What are the results of applying an LPF and HPF to i_1 expressed in Equation (8.17).

8.3.1   Amplitude demodulation

A demodulation process is required to obtain X_1. If we multiply known modulated signal $\sin \omega_c \ t$ with i_1 which is described by Equation (8.16),

$$\begin{aligned} i_1 \cdot \sin \omega_c \ t &= Km \ X_1 \sin(\omega_c \ t + \alpha) \cdot \sin \omega_c \ t \\ &= -\frac{Km}{2} \ X_1[\cos(2\omega_c \ t + \alpha) - \cos \alpha] \quad (8.18) \end{aligned}$$

Equation (8.18) is composed of a high carrier frequency signal, $2\omega_c$, and the constant value $\cos\alpha$. Hence, the first term can be eliminated by applying an LPF. As a result, the low-pass filtered signal $V_1(t)$ is obtained as

$$V_1 = LPF[i_1 \cdot \sin\omega_c\, t] = \frac{Km}{2}\cos\alpha\ X_1 \qquad (8.19)$$

Similarly, i_1 multiplied by $\cos\omega_c\, t$ can be represented as

$$
\begin{aligned}
i_1 \cdot \cos\omega_c\, t &= Km\ X_1\sin(\omega_c\, t + \alpha)\cdot\cos\omega_c\, t \\
&= \frac{Km}{2}[\sin(2\omega_c\, t + \alpha) + \sin\alpha]X_1
\end{aligned}
$$

The first term of $2\omega_c$ can be eliminated by applying an LPF. As a result, the low-pass filtered signal V_2 is obtained as

$$V_2 = LPF[i_1 \cdot \cos\omega_c\, t] = \frac{Km}{2}\sin\alpha\ X_1 \qquad (8.20)$$

To eliminate the α in Equations (8.19) and (8.20), we introduce a trigonometric relation, $\sin^2\alpha + \cos^2\alpha = 1$. Then, a new signal, V^2 is obtained as

$$V^2 = V_1^2 + V_2^2 = \left(\frac{Km}{2}X_1\right)^2 \qquad (8.21)$$

Equation (8.21) reveals that V^2 contains only X_1 without dependence on signal phase delay α. Then, V is obtained by

$$V = \sqrt{\left(\frac{Km}{2}X_1\right)^2} = \frac{Km}{2}X_1 \qquad (8.22)$$

Thus, X_1 is finally demodulated from Equation (8.22) as

$$X_1 = \frac{2}{Km}V$$

The modulated signal $i_1 = Km\ X_1\sin(\omega_c\, t + \alpha)$, can be demodulated using a much simpler method, the root mean square (RMS) method. Firstly, if V is a signal obtained by squaring i_1,

$$
\begin{aligned}
V = i_1^2 &= (Km\ X_1)^2\sin^2(\omega_c\, t + \alpha) \\
&= (Km\ X_1)^2\frac{(1 - \cos 2(\omega_c\, t + \alpha))}{2}
\end{aligned}
$$

Secondly, we take the mean of V, which is approximately equal to the low-pass filtered signal of V, referring to Equation (4.96). Then, we can eliminate the signal of high-carrier frequency $2\omega_c$ to obtain

$$mean[V] = \frac{(Km\,X_1)^2}{2} \tag{8.23}$$

Finally, the root of $mean(V)$ is obtained as

$$\sqrt{mean[V]} = \frac{Km\,X_1}{\sqrt{2}}$$

Hence, X_1 is demodulated as

$$X_1 = \frac{\sqrt{2}}{Km}\sqrt{mean[V]}$$

Example 1: Why is the averaging process of $V(t)$ approximately equal to a low pass filtered signal of $V(t)$?

Solution 1: A periodic function $V(t)$ can be expressed using the Fourier series as

$$V(t) = a_0 + \sum_{m=0}^{\infty}\left(a_k \sin\frac{2\pi kt}{T} + b_k \cos\frac{2\pi kt}{T}\right)$$

$$a_0 = \frac{1}{T}\int_{-\frac{T}{2}}^{\frac{T}{2}} V(t)dt = \frac{1}{T}(\tilde{V}(t)T) = \tilde{V}(t)$$

As represented, $V(t)$ is composed of the DC (constant) value a_0, and signals of high order frequencies. Hence, the low pass filtered signal of $V(t) \cong a_0$. Here, the mathematical expression of a_0 is considered as the mean (or averaging) of $V(t)$. Therefore, it can be said that the mean of $V(t)$ is the same as the low pass filtered signal of $V(t)$.

8.3.2 Amplitude modulation applications in real systems

We studied th control of a contact-type AFM to measure the topology of a material surface in Section 6.9.2. When there are abrupt changes in the height of a material, the fragile cantilever can be easily broken. It is preferable to measure topology without making contact with the sample surface to avoid failure due to mechanical damage. The attractive atomic force between the cantilever tip and material surface is understood as the force generated in a cantilever modeled using a second-order mechanical system composed of cantilever mass m and spring stiffness k, as shown in Figure 8.13. k_1 and k_2 are spring stiffnesses modeled to represent the cantilever which is close to the surface and far from the surface, respectively. Hence, when the cantilever is excited by a frequency near its natural frequency, it vibrates with a large amplitude.

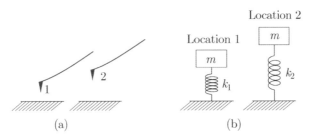

Figure 8.13: (a) A cantilever at different locations in the air and (b) schematic models of a cantilever at different locations in the air.

The oscillation magnitude of the cantilever at different locations can be understood using its frequency response, as shown in Figure 8.14. Suppose the cantilever is excited by a PZT actuator driven at a frequency, ω_m near its natural frequency, ω_s determined at a smaller gap 1. The magnitude of oscillation is obtained from the magnitude ratio response of the cantilever. When the gap increases to have a larger air gap 2, its natural frequency increases to ω_l because of increased spring stiffness. Hence, the magnitude of the cantilever oscillation increases because the magnitude ratio response has a higher slope than that at a smaller gap 1.

The excitation of the cantilever can be understood by mechanical modulation. When a driving voltage of the PZT actuator is described as

$$V(t) = V \sin \omega_m t \tag{8.24}$$

Then, the cantilever displacement $z(t)$ is modulated as

$$z(t) = S(t) \sin(\omega_m t + \theta) \tag{8.25}$$

Here, $S(t)$ is the amplitude which is dependent on the gap magnitude. Figure

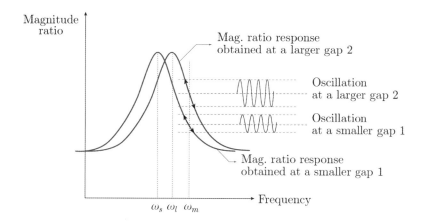

Figure 8.14: Oscillation magnitudes of a cantilever at different locations.

8.15 shows how $z(t)$ varies depending on $S(t)$. Note that Equation (8.25) is the same expression as in the optical amplitude modulation expressed in Equation (8.16). Particularly, amplitude modulation can be realized mechanically to take advantage of the disturbance reduction. The next step is to obtain $S(t)$ by using a demodulation technique.

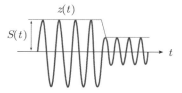

Figure 8.15: $z(t)$ variation depending on gap magnitude $S(t)$.

We use a similar demodulation technique procedure as described in the previous work. If $z(t)$ is multiplied by $V(t)$,

$$V(t)z(t) = VS(t)\sin(\omega_m\ t)\sin(\omega_m t + \theta)$$
$$= -\frac{1}{2}VS(t)[\cos(2\omega_m\ t + \theta) - \cos(-\theta)]$$
$$= \frac{1}{2}VS(t)[-\cos(2\omega_m t + \theta) + \cos\theta]$$

The carrier frequency ω_m can be eliminated by applying an LPF to the above signal. Then, the low pass filtered signal $V_1(t)$ is

$$V_1(t) = LPF[V_1(t)z(t)] = \frac{1}{2}VS(t)\cos\theta \tag{8.26}$$

Similarly, when $z(t)$ is multiplied by $90°$ phase shifted signal, $V(t + 90°)$, we have

$$V(t + 90°)z(t) = VS(t)\cos(\omega_m t)\sin(\omega_m t + \theta)$$
$$= -\frac{1}{2}VS(t)[\sin(2\omega_m t + \theta) + \sin\theta]$$

Then, if the low pass filtered signal of $V(t + 90°)z(t)$ is $V_2(t)$,

$$V_2(t) = LPF[V_1(t + 90°)z(t)] = \frac{1}{2}VS(t)\sin\theta \qquad (8.27)$$

We can obtain $S(t)$ by squaring, adding, and raking the square-rooting $V_1(t)$ and $V_2(t)$, respectively, as previously done in Section 8.3.1.

Alternatively, we can obtain $S(t)$ using the RMS method as introduced in Section 8.3.1. Figure 8.16 shows a block diagram for force control of a non-contact mode AFM.

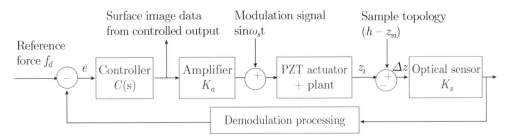

Figure 8.16: Block diagram for force control of a non-contact mode AFM.

Suppose a plant is constructed using a rectangular frame with a hinge-type spring. A cantilever attached to the frame is used for generating the modulated signal. The configuration of the frame and cantilever is shown in Figure 8.17(a). Its schematic diagram is shown in Figure 8.17(b). The rectangular frame is molded using a spring and mass. The cantilever is modeled using a low-stiffness spring and small mass. The resonant frequency of the cantilever is much higher than that of the frame due to its mechanical properties. Because the frequency of the modulated signal is near the natural frequency of the cantilever, it is excited with large oscillation.

The modulated signal can be generated using an additional PZT actuator. However, the modulated high-frequency signal does not affect the mechanical frame because its bandwidth is much lower than the modulated frequency.

Therefore, a single PZT actuator can be used to achieve the purpose of modulated signal generation and force control. Based on the analysis, the modulated signal expressed by Equation (8.24) is added to the amplified control signal and then applied to the plant as shown in Figure 8.16.

The optical sensor measures a low-frequency signal of the controlled output and high-frequency signal of the cantilever oscillation due to the modulated signal. The demodulation process introduced above can be performed by separating the high-frequency signal from the controlled output signal [63] using high-pass or band-pass filters for amplitude measurement. This frequency separation method can be understood by the frequency response analysis introduced in Section 4.4.

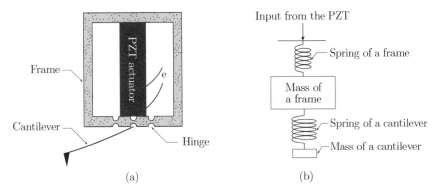

Figure 8.17: (a) A rectangular frame with a hinge type spring and cantilever for non-contact mode and (b) its schematic diagram.

The height of a sample is obtained by measuring the controlled output like the contact mode AFM only if Δz is controlled to be constant as a result of force control. Refer to https://www.ntmdtsi.com/resources/spm-principles/atomic-force-microscopy/amplitude-modulation-afm/non-contactmode to understand its working principle with video demonstrations.

8.4   Phase and frequency modulation

8.4.1   Phase modulation in AFM

In the non-contact mode AFM, the cantilever is placed in the air close to the material surface to avoid contact. When the driving voltage of the PZT actuator is modulated and represented using Equation (8.24), the cantilever oscillation

$z(t)$ is mathematically rewritten as

$$z(t) = S(t)\sin(\omega_m t + \theta(t))$$

Here, $\theta(t)$ is also changed (modulated) as well as $S(t)$ depending on the gap difference. Figure 8.18 shows how $\theta(t)$ varies for different gap magnitudes. $\theta(t)$ has larger variation at a larger gap when the cantilever is modulated at frequency ω_m because the phase delay response has a steeper slope than that at a smaller gap. $\theta(t)$ can be used instead of $S(t)$ because the phase has greater sensitivity than the amplitude.

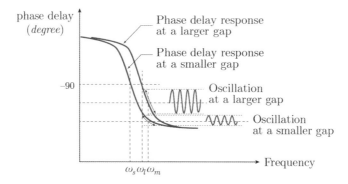

Figure 8.18: Phase variation for different gap magnitudes.

8.4.2 Phase modulation in an optical interferometer

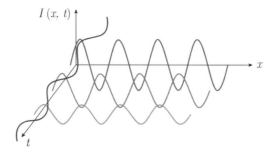

Figure 8.19: Laser beam traveling in the air with respect to time.

The laser beam signal, $I(x, t)$ can be mathematically described in terms of time and space using the wave equation [64] as follows:

$$I(x, t) = E_o \, e^{i(\frac{2\pi}{\lambda}x + \frac{2\pi}{T}t + \phi_o)} \tag{8.28}$$

Here, $\frac{2\pi}{\lambda}$ is spatial wave number, which is simply represented by k; λ and T are the wavelength and time period of the laser beam respectively. E_o is the intensity of the laser beam. Figure 8.19 shows how the laser beam travels in the air with respect to time.

When the space is fixed $I(t) = E_o \cos(\frac{2\pi}{T}t + \phi_o)$. When the time is fixed, $I(x) = E_o \cos(\frac{2\pi}{\lambda}x + \phi_o)$. Hence, λ in the space domain is equivalent to T in the time domain. Figure 8.20 show their mathematical equivalence. The laser

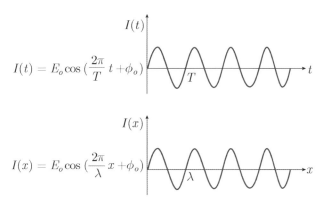

Figure 8.20: Mathematical equivalence in $I(t)$ and $I(x)$

beam can not be measured with a measurement instrument because it has so high frequency, such as terahertz. Instead, an interference effect, using two laser beams for which the interference frequency is lowered, is often used.

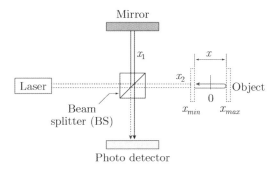

Figure 8.21: An optical system where two laser beams are interfered.

The two laser beams having different optical paths can be interfered with using an optical system, as shown in Figure 8.21. One laser beam is divided into two paths at the beam splitter (BS), which is designed to split the laser beam. The laser beam reflected from the BS goes to the mirror and is transmitted through the beam splitter. This beam is called the reference beam and its path length is denoted by x_1. The laser beam transmitted through the BS goes to the object and is now reflected at the BS. This beam is usually called the received or object beam, and its path length is denoted by x_2. Then, the intensities of the reference beam, $I_1(x,t)$ and received beam $I_2(x,t)$ are respectively represented by

$$
\begin{aligned}
I_1(x,t) &= E_1 e^{i(kx_1 + 2\pi f_1 t + \phi_1)} = E_1 e^{i\epsilon_1} & (8.29) \\
I_2(x,t) &= E_2 e^{i(kx_2 + 2\pi f_2 t + \phi_2)} = E_2 e^{i\epsilon_2} & (8.30)
\end{aligned}
$$

Here, E_1 and E_2 are the amplitudes of the reference beam and received beams, respectively, and $\epsilon_1 = (kx_1 + 2\pi f_1 t + \phi_1)$ and $\epsilon_2 = (kx_2 + 2\pi f_2 t + \phi_2)$.

The intensity of the interference signal, $I(x,t)$ is represented by squaring the summation of $I_1(x,t)$ and $I_2(x,t)$ obtained from Equation (8.29) and Equation (8.30) as

$$
\begin{aligned}
I(x,t) &= |I_1 + I_2|^2 = (I_1 + I_2) \times (I_1 + I_2)^* & (8.31) \\
&= (E_1 e^{i\epsilon_1} + E_2 e^{i\epsilon_2}) \times (E_1 e^{-i\epsilon_1} + E_2 e^{-i\epsilon_2}) \\
&= E_1{}^2 + E_2{}^2 + E_1 E_2 e^{i(\epsilon_1 - \epsilon_2)} + E_1 E_2 e^{-i(\epsilon_1 - \epsilon_2)} \\
&= E_1{}^2 + E_2{}^2 + 2 E_1 E_2 \cos(\epsilon_1 - \epsilon_2)
\end{aligned}
$$

If $\phi_1 = \phi_2$, I is simply represented as

$$
\begin{aligned}
I(x,t) &= E_1{}^2 + E_2{}^2 + 2 E_1 E_2 \cos(k(x_1 - x_2) + 2\pi(f_1 - f_2)t) \\
&= E_1{}^2 + E_2{}^2 + 2 E_1 E_2 \cos(k\Delta x + 2\pi \Delta f t) & (8.32)
\end{aligned}
$$

where $\Delta x = x_1 - x_2$, $\Delta f = f_1 - f_2$.

If $\Delta f = 0$, it is called a homodyne interferometer. Suppose that $E_1^2 = E_2^2 = E_o^2$ for simplicity. When the object moves from x_{min} to x_{max}, $I(x)$ is represented using $\Delta x = 2x$,

$$
I(x) = 2 E_0^2 (1 + \cos 2kx) \tag{8.33}
$$

Note that x is phase modulated. Hence, it can be said that the interferometer is a phase modulator.

Example 2: How is the interference signal $I(x)$ represented for $x = 0, \frac{\lambda}{8}$, and $\frac{\lambda}{4}$, respectively?

Solution 2: According to Equation (8.33), $I = 4E_0^2$ at $x = 0$, $I = 2E_0^2$ at $x = \frac{\lambda}{8}$, and $I = 0$ at $x = \frac{\lambda}{4}$. Figure 8.22 shows how the interfered signal $I(x)$ varies for x. The interfered signal repeats every $\frac{\lambda}{2}$. Therefore, the interferometer can be used for measuring the distance by counting the number of cycles. When He-Ne laser having 780 nm wavelength is used, one cycle of the interfered signal indicates 390 nm distance.

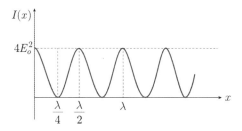

Figure 8.22: Variation of the interference signal $I(x)$ with respect to x.

When there is a Δf, the optical system is called a heterodyne interferometer. A typical example of this configuration is shown in Figure 8.23, which includes optical components of a beam splitter (BS), polarizing beam splitter (PBS), mirror, and acoustic optical modulator (AOM). A polarizing beam splitter (PBS) is used to match the polarization of the reference and the received beam for increasing the S/N of the interference signal [65]. An AOM component is used in the reference beam path for shifting the frequency of the reference beam, f_{laser} by Δf [66]. Hence, the frequency f_{laser} of the reference beam after the AOM is $f_{laser} + \Delta f$. However, the frequency of the received beam f_2 is the same as f_{laser}. Therefore, interfered signal, $I(x, t)$ is represented as

$$
\begin{aligned}
I(x, t) &= 2E_0^2[1 + \cos(2\pi\Delta f t + \frac{2\pi}{\lambda}\Delta x)] \\
&= 2E_0^2[1 + \cos 2\pi(\Delta f t + \frac{2x}{\lambda})] \\
&= 2E_0^2[1 + \cos 2\pi t(\Delta f + \frac{2x}{\lambda t})] \\
&= 2E_0^2[1 + \cos 2\pi t(\Delta f + f_d)] \quad\quad (8.34)
\end{aligned}
$$

Here, f_d is called a Doppler frequency.

$$
f_d = \frac{2x}{\lambda t} = \frac{2v}{\lambda} \quad\quad (8.35)
$$

As expressed in Equation (8.34), the frequency of the interfered signal $I(x,t)$ is $\Delta f + f_d$. As expressed in Equation (8.35), f_d is generated in proportion to the speed of moving object v. Therefore, the heterodyne interferometer is a frequency modulator because the information that we want to know, i.e., v is included in frequency, i.e., f_d.

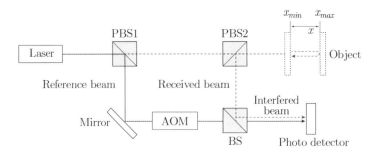

Figure 8.23: Configuration of a heterodyne interferometer.

Example 3: Sketch the heterodyne interference signal when we have displacement motion of $x(t) = \sin(2\pi \cdot 100)t$.

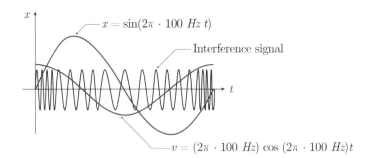

Figure 8.24: Interference signal whose frequency is proportional to the amplitude of $v(t)$.

Solution 3: The velocity $v(t)$ is obtained by differentiating $x(t)$ as

$$v(t) = (2\pi \cdot 100)\cos(2\pi \cdot 100)t$$

Because the interference frequency is proportional to the amplitude of $v(t)$, the interference signal is represented as shown in Fig 8.24.

The frequency shifting effect on f_d by the frequency of Δf can be understood from Figure 8.25. f_d and $-f_d$ are shifted to higher frequencies by Δf. When there is no frequency shifting, it is impossible to measure a negative velocity because negative frequency does not exist. When there is Δf, it is possible to measure a negative velocity because $-f_d$ is shifted to $\Delta f - f_d$, which is the positive frequency range. We can also take advantage of eliminating the disturbances that exist in the low-frequency region using a high-pass filter (HPF).

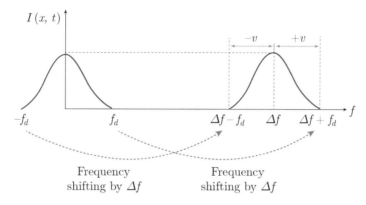

Figure 8.25: Frequency effect on f_d by frequency Δf.

As the velocity of a moving object is proportional to f_d only, the shifting frequency, Δf should be subtracted from the frequency of the interfered signal, $\Delta f + f$, which is called frequency demodulation processing. There are several methods of frequency demodulation introduced in the literature [67],[68].

Q4: Why the velocity signal is preferred for the vibration measurement of mechanical systems compared with the displacement signal?

8.4.3   Phase and frequency modulation in light detection and ranging sensor (LIDAR)

Pulsed LIDAR

Another application example of phase modulation is found in the LIDAR sensor. It is widely used in numerous engineering fields, such as autonomous vehicles, unmanned drones, robot navigation, and 3D scanning applications, owing to its long distance and high-precision measuring capabilities in outdoor environments [69],[70],[71].

There are several LIDAR methods in principle terms. A pulsed time-of-flight (TOF) LIDAR measures distance d to an object by measuring the time difference, Δt, between the reference and received (measured) signals, as shown in Figure 8.26 [72]. Based on the relation of distance velocity relation, d is obtained using constant light velocity c as

$$d = c \times \frac{\Delta t}{2} \qquad (8.36)$$

Δt is divided by two because the laser beam travels twice the distance d.

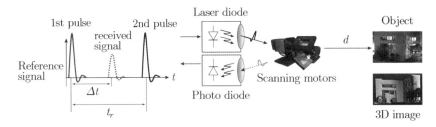

Figure 8.26: Measurement principle of pulsed time-of-flight method.

Δt can be measured using a time-to-digital converter, where the time interval between two signal pulses is counted using the number of pulses. The LIDAR speed is determined by how fast the pulse of the reference signal is emitted from the LD. If the repetition rate of the pulse is defined by t_r, it should be longer than the travel time required to measure the shortest distance. A 3D image is obtained using two scanners rotating in horizontal and vertical directions. When a high-resolution 3D image needs to be constructed, a small t_r or low-speed scanner rotating in horizontal and vertical directions is required. However, when a high-resolution, high-speed 3D image needs to be constructed, high-speed scanners are important specifications in addition to a small t_r.

Amplitude modulated continuous wave (AMCW) LIDAR

In an AMCW LIDAR, the distance can be measured using the phase difference between the modulated reference signal and object signals, instead of the time difference. When light travels distance d, light reflected from an object is delayed by phase φ as shown in Figure 8.27. The phase difference is proportional to the distance traveled and obtained using the following relation:

$$\frac{\varphi}{2\pi} = \frac{2d}{\lambda}$$

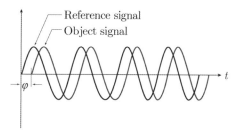

Figure 8.27: Measurement principle of an AMCW LIDAR.

Hence, the distance traveled d can be obtained as

$$d = \frac{\varphi \lambda_m}{4\pi} \quad (8.37)$$

Here, λ_m is the wavelength of the modulated reference signal. If f_m is the frequency of the modulated reference signal, $f_m \lambda_m = c$, where c is the speed of light, $3 \times 10^8 m/s$. the delayed phases are the same regardless of the modulating frequencies. Equation (8.37) is expressed again using f_m as follows:

$$d = \frac{c \, \varphi}{4\pi f_m} \quad (8.38)$$

Hence, a higher modulated frequency provides a better distance resolution.

The mathematical expression of the modulated object signal, $I_m(t)$ is

$$I_m(t) = A_m \cos(f_m t + \varphi) \quad (8.39)$$

If we find the phase difference φ that is proportional to the distance using a demodulation technique, d can be obtained. We can apply the previously introduced demodulation technique. By mixing the reference signal $V_r(t) = A_r cos(f_m t)$ with the modulated received beam signal expressed in Equation (8.39), we obtain

$$V_o(t) = A_r A_m \cos(f_m t) \cos(f_m t + \varphi)$$
$$= \frac{1}{2} A_m A_r [\cos(2f_m t + \varphi) + \cos \varphi]$$

A low pass filtered output of $V_o(t)$ gives the phase as a result. This type of processing is also called a Lock-in amplifier.

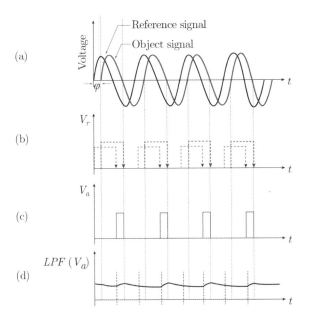

(a)

(b)

(c)

(d)

Figure 8.28: (a) Reference and object signals with the phase difference φ, (b) rectified signal V_r using digital electronics such as Schmitt-trigger, (c) V_a obtained using the falling edges, and (d) low pass filtered signal $LPF[V_a]$

Alternatively, we can find the phase by directly using electronic circuits. When we obtain a reference and an object signal with the phase difference φ, as shown in Figure 8.28(a), the rectified signals of the reference and an object signal, V_r can be obtained using digital electronics, such as Schmitt-trigger as shown in Figure 8.28(b). Using the falling edges of the two signals, we can obtain a new signal V_a, as shown in Figure 8.28(c). When V_a is filtered using an LPF with a lower cutoff frequency, the pseudo-DC (constant) value will be obtained as shown in Figure 8.28(d) [73]. The larger the φ, the larger the magnitude of the DC value is obtained because the area is proportional to φ.

Q5: How is the LPF cutoff frequency selected in order to make the low-pass filtered signal correspond to the phase difference?

An AMCW LIDAR sensor has a higher distance resolution than TOF LIDAR [74]. This advantage is due to a modulation technique that can be easily applied to the continuous wave used in AMCW LIDAR. To obtain higher distance resolution AMCW, it is preferable to increase the modulation frequency, as expressed in Equation (8.38). However, the received signal at a high modulation frequency is easily contaminated with electronic noise during signal processing of the photo detector when converting to voltage. This noise can jitter the received signal. Therefore, if the phase is demodulated directly from the high modulation frequency, the jitter problem can yield an inaccurate phase measurement. This can be solved by shifting the high modulation frequency to a lower frequency during the demodulation process. For example, if a modulated reference fre-

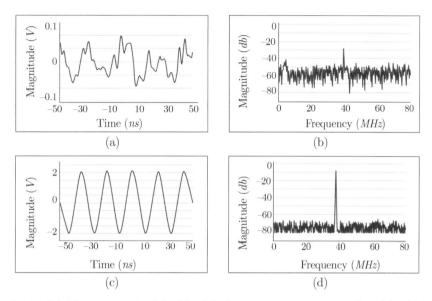

Figure 8.29: (a) Time signal of $I_m(t)$, (b) frequency spectrum of $I_m(t)$, (c) time signal of the intermediate signal $I_i(t)$, and (d) frequency spectrum of $I_i(t)$.

quency of 40 MHz is applied, the detecting signal, $I_m(t)$, in the detector is contaminated with noise $N(t)$, as shown in Figure 8.29(a). The frequency spectrum of $I_m(t)$ is shown in Figure 8.29(b) and its magnitude is not that large compared with the noise spectrum, indicating a very low signal-to-noise ratio (S/N ratio). According to Equation (8.39), the mathematical description of $I_m(t)$ is

$$I_m(t) = A_m \cos(40MHz\ t + \varphi) + N(t)$$

To increase the S/N ratio, if $I_m(t)$ is mixed with 39.95 MHz signal of amplitude A_r, we obtain the intermediate output $I_i(t)$ using the trigonometric relation as

$$
\begin{aligned}
I_i(t) &= A_m A_r \cos(39.95MHz\ t)(\cos(40MHz\ t + \varphi) + N(t)) \\
&= \frac{1}{2} A_m A_r [\cos(79.95MHz\ t + \varphi) + \cos(50kHz\ t + \varphi)] \\
&\quad + A_m A_r \cos(39.95MHz\ t)N(t)
\end{aligned}
$$

Then, the low-pass filtered signal of $I_i(t)$ is $\frac{1}{2}A_m A_r \cos(50kHz\ t + \varphi)$, as shown in Figure 8.29(c). It is obvious that $I_i(t)$ is quite clean. The frequency spectrum of $I_i(t)$ is shown in Figure 8.29(d), where its magnitude is large compared to the noise spectrum.

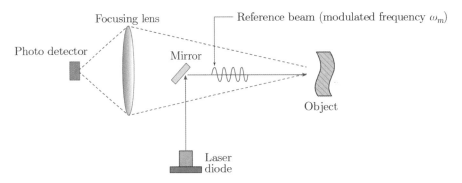

Figure 8.30: System configuration of the AMCW LIDAR.

One of the AMCW LIDAR configurations is shown in Figure 8.30. Figures 8.31(a) and (b) show the real image of an object and its 3D scanning image obtained using the phase demodulation method introduced above.

(a)   (b)

Figure 8.31: (a) Actual image of an AMCW LIDAR and (b) 3D scanning result obtained by an AMCW LIDAR using the phase demodulation method.

Frequency-modulated continuous wave (FMCW) LIDAR

The distance can be also measured using the frequency modulation method. This is called a FMCW LIDAR. It has gained attention lately because it is very robust in harsh environmental conditions and sunlight compared to pulsed TOF LIDAR. This is because FMCW LIDAR takes advantage of the wave properties of the laser. Hence, the laser beam can propagate through rain, snow, and fog. Additionally, FMCW LIDAR can measure the speed of an object by the Doppler effect as well as the distance to an object [75],[76].

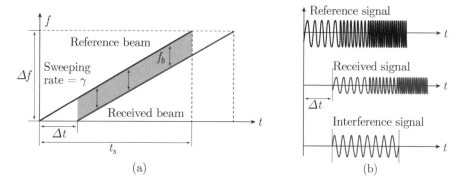

Figure 8.32: (a) Modulated signals of the reference beam and received beam and (b) time signals of the reference, received, and interfered beams.

Figure 8.32(a) shows how to measure distance in FMCW LIDAR. When the frequency of the reference beam is linearly modulated up to frequency Δf during the sweeping time t_s, the reflected beam delayed by time difference Δt is received by an optical detector. Figure 8.32(b) shows the time signals of the reference and received beams. Here, t is proportional to the distance to the object, as discussed for the TOF LIDAR. When the beams are interfered with, an interference signal with frequency f_b is obtained. Then, Δt is indirectly obtained by f_b.

To know the relation of distance d between LIDAR and the object, the above the process is mathematically represented. First, the chirp rate of the linear frequency modulation γ defined as $\frac{\Delta f}{t_s}$ can be also represented by f_b and Δt as

$$\gamma = \frac{f_b}{\Delta t} \tag{8.40}$$

Then, d can be rewritten by substituting Equation (8.40) into Equation (8.36) as

$$d = \frac{c \cdot \Delta t}{2} = \frac{c \cdot f_b}{2\gamma} \tag{8.41}$$

Thus, d can be measured using f_b instead of Δt, which differs from conventional pulsed TOF LIDAR. The distance resolution can also be determined using Equation (8.41). The larger the γ used, the smaller the obtained d. This can also be graphically understood from Figure 8.32(a), which shows that a larger f_b is obtained for the same t_s when γ is increased.

Let's examine how the reference beam and received beam are interfered with and theoretically represented. Suppose that the intensity of the reference beam, $I_1(x, t)$, is represented by

$$I_1(x, t) = E_1 \cos(f_1 t + \phi_1) \tag{8.42}$$

where E_1, f_1, and ϕ_1 are the amplitude, frequency, and phase of $I_1(x, t)$, respectively. The received beam is delayed by Δt, then, the intensity of the received beam, $I_2(x, t)$ is represented by

$$I_2(x, t) = E_2 \cos((f_1 - \gamma \Delta t)t + \phi_2) \tag{8.43}$$

where E_2 and ϕ_2 are the amplitude and phase of $I_2(x, t)$, respectively. Then, the intensity of the interfered beam, I is obtained by squaring the summation of $I_1(x, t)$ and $I_2(x, t)$ referring to Equation (8.31) as

$$I = E_1{}^2 + E_2{}^2 + 2E_1 E_2 \cos(f_1 t + \phi_1) \cdot \cos((f_1 - \gamma \Delta t)t + \phi_2) \tag{8.44}$$

Equation (8.44) can be rearranged using the trigonometric relation as follows:

$$\begin{aligned} I = {} & E_1{}^2 + E_2{}^2 + E_1 E_2 \left(\cos((2f_1 - \gamma \Delta t)t + \phi_1 + \phi_2) \right. \\ & \left. + \cos((\gamma \Delta t)t + \phi_1 - \phi_2) \right) \end{aligned} \tag{8.45}$$

The DC (constant) offset in Equation (8.45) can be easily eliminated using a high pass filter. As the first cosine-term has terahertz frequency, it cannot be measured using a photodetector. Therefore, only the second cosine term is measured and a final interfered beam signal, I_s, can be obtained under the assumptions of $\phi_1 - \phi_2 = 0$ and using Equation (8.40) as

$$I_s = \cos(f_b t) \tag{8.46}$$

f_b can be demodulated using a frequency detector.

There is a device called a tunable laser where the wavelength of a laser increases proportionally with increasing current [77]. Hence, it can be used to modulate the frequency of a laser beam, using the principle that the frequency of a laser is inversely proportional to the wavelength of the laser. Though current can be used for varying the frequency of a laser, it is almost impossible to achieve linear frequency modulation by just applying current to a tunable laser. As a result, when the current is increased proportionally, f_b actually changes at the same distance with respect to time, as shown in Figure 8.33(a). This causes distance measurement inaccuracy. A large variance of f_b is observed with a wide frequency spectrum, Δf_b, as shown in Figure 8.33(b). Hence, linear frequency-modulating control of the tunable laser is required

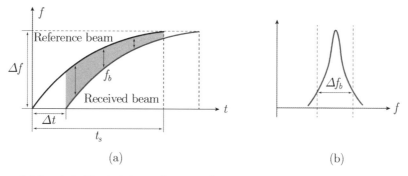

(a)   (b)

Figure 8.33: (a) Varied interference frequency f_b at the same distance with respect to time and (b) frequency spectrum of Δf_b.

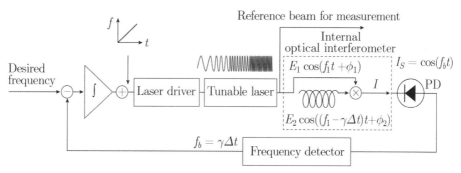

Figure 8.34: Frequency-locked loop of FMCW LIDAR.

To achieve linear frequency-modulating control, the frequency of the interference signal at a known distance is controlled using an additional optical interferometer installed internally, as shown in Figure 8.34. When the received beam traveling a known distance is interfered with the reference beam, the corresponding interference frequency can be calculated. This frequency is a desired frequency f_d for feedback control. The tunable laser is a plant to be controlled. A frequency detector is used for calculating f_b as a sensor, and the integrator is used as a controller. This frequency controller is usually called a frequency-locked loop (FLL). If the frequency of the tunable laser is well controlled, the frequency of the reference beam can be linearly modulated using a ramp input added to the tunable laser. Then, the linearly modulated reference beam can be emitted onto an object for measurement.

Figure 8.35(a) shows the time response of the output voltage, V_{int} corresponding to the interfered signal, I_s represented by Equation (8.46). Figure 8.35(b) shows the time response of the sensor output voltage, V_f corresponding to interfered frequency f_b, which can be realized using the electronic component, frequency–voltage converter (FVC) [67]. It is observed that V_f is quickly controlled by f_d to reach a steady-state condition within 0.1 ms. Figure 8.35(c) shows the frequency spectrum of f_b obtained from the linear frequency-sweeping control. It is observed that the frequency spectrum is very narrow as expected. However, the frequency spectrum becomes very wide when the linear frequency sweeping control is not used as shown in Figure 8.36.

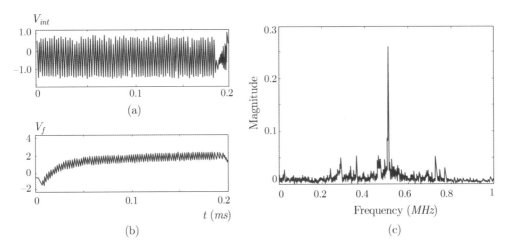

Figure 8.35: Time responses of (a) V_{int} corresponding to the interfered signal, I_s, (b) V_f corresponding to the interfered frequency, f_b, and (c) narrow frequency spectrum obtained from the linear frequency sweeping control.

When the linear frequency modulation is successfully implemented, we can scan the tunable laser as a reference beam for the distance measurement. Figure 8.37 shows a real image of an object and its 2D scanning image obtained using the Fast Fourier Transformation (FFT) method for frequency demodulation.

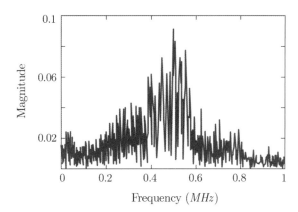

Figure 8.36: Wide frequency spectrum obtained without the linear frequency sweeping control.

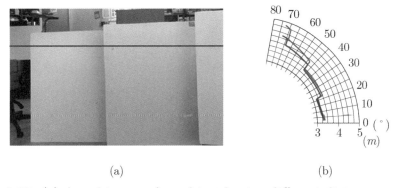

Figure 8.37: (a) A real image of an object having different distances and (b) its 2D scanning image using the Fast Fourier Transformation method.

Bibliography

[1] https://www.123rf.com/photo 32442180 male worker operating excavator on construction site. html.

[2] Principle, design, and future of inchworm type piezoelectric actuators, Li, Jianping, et al., Intechopen, ISBN: 9781839688317 (1839688319), 2021.

[3] Introduction to Physical System Dynamics, Ronald C. Rosenbert, Dean C. Karnopp, McGraw-Hill Book Company.

[4] Modern Control Engineering, Katsuhiko Ogata, Prentice-Hall International, second edition.

[5] System Dynamics and Control with Bond Graph Modeling, Javier Kypuros, CRC press, ISBN 978-1466560758, 2013.

[6] Bond graphs for engineers, edited by P.C. Breedveld, G. Dauphin-Tanguy, from IMACS 13th World Congress, Dublin, Ireland, July 1991.

[7] Vector mechanics for engineers, Dynamics, third edition, Ferdinand P, Beer. E., Russel Johnston Jr., McGraw-Hill Book Company.

[8] DORF, R. C., SVOBODA, J. A. (2013), Introduction to electric circuits, Hoboken, NJ, J. Wiley Sons. ISBN: 9781118477502.

[9] Svelto O., Hanna D. C., Principles of lasers (Vol. 4), New York: Plenum Press, 1998.

[10] Munson, Bruce Roy, et al., Fluid Mechanics, Singapore: Wiley, 2013.

[11] Reuben M. Olson, Essentials of Engineering Fluid Mechanics, Fourth edition, Harper Row.

[12] Minjae Cho, Eunsung Kwon, Kyihwan Park, "Design and Control of Pneumatic System for Recycling Classification of Non-Ferrous Metals," International Journal of Precision Engineering and Manufacturing-Green Technology, Vol 9.2, pp 567-575, 2022.

[13] Erwin Kreyszig, Advanced Engineering Mathematics, Wiley International edition.

[14] Karnopp, Dean, Donald L. Margolis, and Ronald Carl Rosenberg, System Dynamics, New York: Wiley, 1990.

[15] Preumont, André, Vibration control of active structures: an introduction, Vol. 246. Springer, 2018.

[16] Jongkyu Jung, Woosub Youm, SungQ Lee, Kyihwan Park, "Vibration reduction control of a voice coil motor(VCM)-driven actuator for SPM applications," INT J ADV. MANUF. TECH, 30 April 2009.

[17] Herbert H. Woodson, James R Melcher, Electromechanical Dynamics, Part 1: Discrete Systems, Krieger publishing company, Florida, 1990.

[18] K. Lee, Kyihwan Park, "Modeling of eddy currents with boundary conditions by using Coulomb's law and the method of image," IEEE Transactions on Magnetics, Vol. 38, No. 2, pp. 1333-1340, March 2002.

[19] S. Wang, J. Kang, Kyihwan Park, H.S. Yoon, G.H.Jang, "Comparison of 2D and 3D FEA of BLDC motor," COMPEL, Vol. 19, No 2, pp.529-537, 2000.

[20] Introduction to Electrodynamics, David J. Griffiths, 2nd edition, Prentice Hall.

[21] William H. Hayt, Jr., Engineering Electromagnetics, McGraw-Hill Book Company.

[22] Halliday, David; Resnick, Robert; Walker, Jearl: Fundamentals of Physics, 2013.

[23] Riley, Chris, Effect of Magnetic Hysteresis in Solenoid Valve Operation, Sensor Letters. 11. 9-12. 10.1166/s, 2013.

[24] Hanselman Duane, et al. Brushless Motors: Magnetic Design, Performance, and Control of Brushless DC and Permanent Magnet Synchronous Motors, E-Man Press LLC, 2012.

[25] Büchi, Roland. Brushless Motors and Controllers: Books On Demand, 2012.

[26] Yongdae Kim, Sangyoo Kim, and Kyihwan Park, "Magnetic force driven six degree-of-freedom active vibration isolation system using a phase compensated velocity sensor," Review of Scientific Instruments, Vol 80, 045108-1, pp. 1 ~ 5, 23 April 2009.

[27] Jongkyu Jung, Woosub Youm, SungQ Lee, Kyihwan Park, "Vibration reduction control of a voice coil motor (VCM)-driven actuator for SPM applications," Int. J. Adv. Manuf. Tech., 30 April 2009.

[28] Woosub Youm, Jongkyu Jung, Sung-Q Lee, Kyihwan Park, "Control of voice coil motor nanoscanners for an atomic force microscopy system using a loop shaping technique," Review of Scientific Instruments, Vol. 79, 013707-1 pp. 1 ~ 6, January 17, 2008.

[29] Kyihwan Park, K.Y. Ahn, S.H. Kim, Y.K. Kwak, "Wafer Distribution System for a Clean Room Using a Novel Magnetic Suspension System Technique," IEEE/ASME Trans. on Mechatronics, Vol. 3, No. 1, pp. 73-78, 1998.

[30] http://www.em4sys.co.kr/main/.

[31] Kapjin Lee, Kyihwan Park, "Analysis of an eddy-current brake considering the finite radius and induced magnetic flux," J. of Applied Physics, 2002, Vol.92, No.9, pp.5532-5538, 1 November 2002.

[32] K. Lee, Kyihwan Park, "Modeling of eddy currents with boundary conditions by using Coulomb's law and the method of image," IEEE Transactions on Magnetics, Vol. 38, No. 2, pp. 1333-1340, March 2002.

[33] Shabana, Ahmed A., Dynamics of Multi-body Systems, Cambridge University Press, 2003.

[34] A. G. Chassiakos and G. A. Bekey, "On the modeling and control of a flexible manipulator arm by point actuators," 25th IEEE Conference on Decision and Control, pp. 1145-1150, doi: 10.1109/CDC.267561, 1986.

[35] Yoram Koren, Robotics for Engineers, McGraw-Hill International Editions, ISBN: 007100534X, 1987.

[36] Baccouch M, Dodds S, "A Two-Link Robot Manipulator: Simulation and Control Design," International Journal of Robotic Engineering Vol. 5, Issue 2, 2020, ISSN 2631-5106, DOI: 10.35840/2631-5106/4128.

[37] Modern Control Engineering, Katsuhiko Ogata, Prentice-Hall International, Second edition.

[38] J. H. Yi, Kyihwan Park, S. H. Kim, Y. K. Kwak, M. Abdelfatah, and I. Busch-Vishniac, "Robust Force Control for a Magnetically Levitated Manipulator Using Flux Density Measurement," A Journal of IFAC, Vol. 4, No. 7, pp.957-965, 1996.

[39] Suomalainen, M., Karayiannidis, Y., Kyrki, V., A survey of robot manipulation in contact, Robotics and Autonomous Systems, 156, 104224. https://doi.org/10.1016/j.robot.2022.104224.

[40] R.C. Dorf, Modern Control Systems, 6th edition, Addison-Wesley.

[41] Nise, N. S., Control Systems Engineering, 6th edition, Wiley, 2010.

[42] Steven K. Thompson, Sampling, J. Wiley Sons. ISBN: 9780470402313, 2012.

[43] Gruyitch, Lyubomir. Tracking Control of Linear Systems, ISBN: 9780367379995, 2016.

[44] Gene F. Franklin, J. David Powell, Michael L. Workman, Digital Control of Dynamic Systems, 2nd edition, Addison-Wesley publishing company, ISBN: 0201518848.

[45] Li Tan Jean Jiang, Digital Signal Processing: Fundamentals and Applications. Academic Press. ISBN: 9780124158931, 2013.

[46] Kyihwan Park, K. B. Choi, S. H. Kim, Y. K. Kwak, "Magnetic Levitated High Precision Positioning System Based on Antagonistic Mechanism," IEEE Transactions on Magnetics, Vol. 32, No. 1, pp.208-219, 1996.

[47] J.M. Maciejowski, Multivariable Feedback Design, Addison-Wesley publishing company, ISBN: 0201182432.

[48] Chulsoo Kim, Jongkyu Jung, and Kyihwan Park, "Vibration reduction control of an atomic force microscope using an additional cantilever," Review of Scientific Instruments, 82, 116102, Nov 2011.

[49] Stephan J. G. Gift, Brent Maundy. (2021). Electronic Circuit Design and Application, Springer Cham., ISBN : 9783030469917.

[50] Yongdae Kim, Sangyoo Kim, and Kyihwan Park, "Magnetic force driven six degree-of-freedom active vibration isolation system using a phase compensated velocity sensor," Review of Scientific Instruments, Vol 80, pp045108-1 5, 23 April 2009.

[51] Woosub Youm, Jongkyu Jung, Sung-Q Lee, Kyihwan Park, "Control of voice coil motor nanoscanners for an atomic force microscopy system using a loop shaping technique," Review of Scientific Instruments, Vol 79, pp.013707-1 6, 17 January 2008.

[52] Stephan J. G. Gift, Brent Maundy. (2021). Electronic Circuit Design and Application, Springer Cham., ISBN: 9783030469917.

[53] Alexander, C. K., Sadiku, M. N. O., Fundamentals of Electric Circuits, Boston: McGraw-Hill Higher Education, ISBN: 9780071284417, 2017.

[54] https://www.analog.com/media/en/technical-documentation/data-sheets/AD8021.pdf.

[55] https://fritzing.org.

[56] Weiss, Mark Allen, and Yue Chen, Data Structures and Algorithm Analysis in C., California: Benjamin/Cummings, 1993.

[57] Buffolo, M., De Santi et al. M., "A Review of the Reliability of Integrated IR Laser Diodes for Silicon Photonics," Electronics 10, 2734, 2021.

[58] Robert L. Boylestad and Louis Nashelsky, Electronic Devices and Circuit Theory, Pearson, ISBN: 978-9332542600, 2015.

[59] Pearsall, Thomas, Photonics Essentials, 2nd edition. McGraw-Hill. ISBN 978-0-07-162935-5, 2010.

[60] Jongpil La, Kyihwan Park, "Signal Processing Algorithm of a Position Sensitive Detector(PSD) using Amplitude Modulation/Demodulation," Review of Scientific Instruments, Vol. 76, No. 2, pp.24701-24706, January 2005.

[61] J. F. James, A student's Guide to Fourier Transformations, Cambridge University Press, ISBN: 0521462983.

[62] Jong Kyu Jung, Seong gu Kang, Joon Sik Nam, and Kyi Hwan Park, "Intensity control of triangulation based PSD sensor independent of object color variation," Sensors Journal, IEEE, Vol. 11, No. 12, pp.3311-3315, 2011.

[63] Sung-Q Lee, Woo-sub Woum, Ki-Bong Song, Eun-Kyung Kim, Jun-Ho Kim, Kyi-Hwan Park, Kang-Ho Kang, "Frequency response separation scheme for non-contact type atomic force microscope," Sensors and Actuators A, Vol. 116, pp.45-50, May 2004.

[64] Daniel Fleisch, Laura Kinnaman, A Student's Guide to Waves (Student's Guides), 1st Edition, Cambridge University Press, ISBN: 978-1107643260, 2015.

[65] Serge Huard, Polarization of Light, New York: John Wiley Sons, 1997.

[66] SHAMIR, Joseph. Optical Systems and Processes, SPIE Press, 1999.

[67] Jongpil La, Hyunseung Choi, and Kyihwan Park, "Heterodyne laser Doppler vibrometer using a Zeeman-stabilized He–Ne laser with a one-shot frequency to voltage converter," Review of Scientific Instruments, Vol. 76, No. 2, pp. 25112-25118, January 2005.

[68] Seonggu Kang, Jongpil La, Heesun Yoon, and Kyihwan Park, "A synthetic heterodyne interferometer for small amplitude of vibration measurement," Review of Scientific Instruments, Vol. 79, pp. 053106-1 6, 30 May 2008.

[69] T. Ogawa, H. Sakai, Y. Suzuki, K. Takagi, and K. Morikawa, "Pedestrian detection and tracking using in-vehicle lidar for automotive application," in 2011 IEEE Intelligent Vehicles Symposium (IV), pp. 734– 739, 2011.

[70] J. Jang, S. Hwang, and K. Park, "Note: Optical and electronic design of an amplitude-modulated continuous-wave laser scanner for high accuracy distance measurement," Review of Scientific Instruments, Vol. 86, No. 4, 046104, 2015.

[71] U. Weiss and P. Biber, "Plant detection and mapping for agricultural robots using a 3D Lidar sensor," Robotics and Autonomous Systems, Vol. 59, No. 5, pp. 265–273, 2011.

[72] Y. Cho, Y. Yoon, J. Lyu, and Kyihwan Park, "Auto gain control method using the current sensing amplifier to compensate the walk error of the TOF lidar," 19th International Conference on Control, Automation and Systems (ICCAS), pp. 1403–1406, 2019.

[73] Sungui Hwang, Junhwan Jang, and Kyihwan Park, "Continuous-wave time-of-flight laser scanner using two laser diodes to avoid 2π ambiguity," Review of Scientific Instruments, 84, 086110, August 2013.

[74] Heesun Yoon, Hajun Song, and Kyihwan Park, "A phase-shift laser scanner based on time counting method for high linearity performance," Review of Scientific Instruments, 82, 075108, July 2011.

[75] M. Kutila, P. Pyykönen, W. Ritter, O. Sawade, and B. Schäufele, "Automotive lidar sensor development scenarios for harsh weather conditions," in IEEE 19th International Conference on Intelligent Transportation Systems (ITSC), pp. 265–270, 2016.

[76] M. Kutila, P. Pyykönen, H. Holzhüter, M. Colomb, and P. Duthon, "Automotive lidar performance verification in fog and rain," 21st International Conference on Intelligent Transportation Systems (ITSC), pp. 1695–1701, IEEE, 2018.

[77] K. Iiyama, L.-T. Wang, and K.-I. Hayashi, "Linearizing optical frequency-sweep of a laser diode for fmcw reflectometry," Journal of Lightwave Technology, Vol. 14, No. 2, pp. 173–178, 1996.

Modeling and Control
of Physical Systems

초 판 인 쇄 2023년 10월 13일
초 판 발 행 2023년 10월 20일

지 은 이 Kyihwan Park(박기환)
발 행 인 임기철
발 행 처 GIST PRESS

등 록 번 호 제2013-000021호
주 소 광주광역시 북구 첨단과기로 123(오룡동)
대 표 전 화 062-715-2960
팩 스 번 호 062-715-2069
홈 페 이 지 https://press.gist.ac.kr/
인쇄 및 보급처 도서출판 씨아이알(Tel. 02-2275-8603)

I S B N 979-11-90961-18-9 (93550)
정 가 25,000원